CRJ 900 Company Flight Manual Volume 2

Printed in U.S.A.

Intentionally Left Blank

Record of Revisions

Rev. No.	Effective Date
0	01 Nov 2013
1	14 Nov 2014
2	21 Jul 2015
3	28 Oct 2015
4	
5	
6	
7	
8	
9	
10	
11	
12	
13	
14	
15	

Rev. No.	Effective Date
16	
17	
18	
19	
20	
21	
22	
23	
24	
25	
26	
27	
28	
29	
30	
31	

Rev. No.	Effective Date
32	
33	
34	
35	
36	
37	
38	
39	
40	
41	
42	
43	
44	
45	
46	
47	

Intentionally Left Blank

List of Bulletins

Bulletin No.	Effective Date	Subject	Filing Instructions	Date Removed
15-01	17 Sep 15	FMS		28 Oct 15

Intentionally Left Blank

List of Temporary Revisions

TR No.	Effective Date	Subject	Filing Instructions	Date Removed
15-01	25 Mar	Updated Engine start to recommend FO start engines. Update holding reference, and new TLR	remove and replace pages	28 OCT 15 With rev 3

Intentionally Left Blank

List of Effective Pages

Page No.	Revision No.	Effective Date
Record of Revisions		
i	03	28 Oct 2015
ii	0	01 Nov 2013
List of Bulletins		
iii	0	01 Nov 2013
iv	0	01 Nov 2013
List of Temporary Revisions		
v	03	28 Oct 2015
vi	0	01 Nov 2013
List of Effective Pages		
vii	03	28 Oct 2015
viii	03	28 Oct 2015
ix	03	28 Oct 2015
x	03	28 Oct 2015
xi	03	28 Oct 2015
xii	03	28 Oct 2015
xiii	03	28 Oct 2015
xiv	03	28 Oct 2015
xv	03	28 Oct 2015
xvi	03	28 Oct 2015
xvii	03	28 Oct 2015
xviii	03	28 Oct 2015
xix	03	28 Oct 2015
xx	03	28 Oct 2015
xxi	03	28 Oct 2015
xxii	3	28 Oct 15
Master Table of Contents		
xxiii	3	28 Oct 15
xxiv	3	28 Oct 15
xxv	3	28 Oct 15
xxvi	3	28 Oct 15

Page No.	Revision No.	Effective Date
Preface		
xxvii	03	28 Oct 2015
xxviii	03	28 Oct 2015
Chapter 1 TOC		
i	2	21 Jul 2015
ii	2	21 Jul 2015
Chapter 1		
1	2	21 Jul 15
2	2	21 Jul 15
Chapter 2 TOC		
i	01	14 Nov 2014
ii	01	14 Nov 2014
iii	01	14 Nov 2014
iv	0	01 Nov 2013
Chapter 2		
1	01	14 Nov 2014
2	01	14 Nov 2014
3	01	14 Nov 2014
4	01	14 Nov 2014
5	01	14 Nov 2014
6	01	14 Nov 2014
7	01	14 Nov 2014
8	01	14 Nov 2014
9	01	14 Nov 2014
10	01	14 Nov 2014
11	01	14 Nov 2014
12	01	14 Nov 2014
13	01	14 Nov 2014
14	01	14 Nov 2014
15	01	14 Nov 2014
16	01	14 Nov 2014
17	01	14 Nov 2014
18	01	14 Nov 2014

Page No.	Revision No.	Effective Date
19	01	14 Nov 2014
20	01	14 Nov 2014
21	01	14 Nov 2014
22	01	14 Nov 2014
23	01	14 Nov 2014
24	01	14 Nov 2014
25	01	14 Nov 2014
26	01	14 Nov 2014
27	01	14 Nov 2014
28	01	14 Nov 2014
29	01	14 Nov 2014
30	01	14 Nov 2014
31	01	14 Nov 2014
32	01	14 Nov 2014
33	01	14 Nov 2014
34	01	14 Nov 2014
35	01	14 Nov 2014
36	01	14 Nov 2014
37	01	14 Nov 2014
38	01	14 Nov 2014
39	01	14 Nov 2014
40	01	14 Nov 2014
41	01	14 Nov 2014
42	01	14 Nov 2014
43	01	14 Nov 2014
44	01	14 Nov 2014
45	01	14 Nov 2014
46	01	14 Nov 2014
47	01	14 Nov 2014
48	01	14 Nov 2014
49	01	14 Nov 2014
50	01	14 Nov 2014

Chapter 3 TOC

Page No.	Revision No.	Effective Date
i	03	28 Oct 2015
ii	03	28 Oct 2015
iii	03	28 Oct 2015
iv	03	28 Oct 2015

Chapter 3

Page No.	Revision No.	Effective Date
1	01	14 Nov 2014
2	0	01 Nov 2013
3	0	01 Nov 2013
4	02	21 Jul 2015
5	03	28 Oct 2015
6	02	21 Jul 2015
7	02	21 Jul 2015
8	02	21 Jul 2015
9	02	21 Jul 2015
10	02	21 Jul 2015
11	02	21 Jul 2015
12	02	21 Jul 2015
13	02	21 Jul 2015
14	3	28 Oct 2015
15	02	21 Jul 2015
16	02	21 Jul 2015
17	02	21 Jul 2015
18	03	28 Oct 2015
19	03	28 Oct 2015
20	03	28 Oct 2015
21	03	28 Oct 2015
22	03	28 Oct 2015
23	03	28 Oct 2015
24	03	28 Oct 2015
25	03	28 Oct 2015
26	03	28 Oct 2015
27	03	28 Oct 2015
28	03	28 Oct 2015
29	03	28 Oct 2015
30	03	28 Oct 2015
31	03	28 Oct 2015
32	03	28 Oct 2015

Page No.	Revision No.	Effective Date	Page No.	Revision No.	Effective Date
33	03	28 Oct 2015	69	03	28 Oct 2015
34	03	28 Oct 2015	70	03	28 Oct 2015
35	03	28 Oct 2015	71	03	28 Oct 2015
36	03	28 Oct 2015	72	03	28 Oct 2015
37	03	28 Oct 2015	73	03	28 Oct 2015
38	03	28 Oct 2015	74	03	28 Oct 2015
39	03	28 Oct 2015	75	03	28 Oct 2015
40	03	28 Oct 2015	76	03	28 Oct 2015
41	03	28 Oct 2015	77	03	28 Oct 2015
42	03	28 Oct 2015	78	03	28 Oct 2015
43	03	28 Oct 2015	79	03	28 Oct 2015
44	03	28 Oct 2015	80	03	28 Oct 2015
45	03	28 Oct 2015	81	03	28 Oct 2015
46	03	28 Oct 2015	82	03	28 Oct 2015
47	03	28 Oct 2015	83	03	28 Oct 2015
48	03	28 Oct 2015	84	03	28 Oct 2015
49	03	28 Oct 2015	85	03	28 Oct 2015
50	03	28 Oct 2015	86	03	28 Oct 2015
51	03	28 Oct 2015	87	03	28 Oct 2015
52	03	28 Oct 2015	88	03	28 Oct 2015
53	03	28 Oct 2015	89	03	28 Oct 2015
54	03	28 Oct 2015	90	03	28 Oct 2015
55	03	28 Oct 2015	91	03	28 Oct 2015
56	03	28 Oct 2015	92	03	28 Oct 2015
57	03	28 Oct 2015	93	03	28 Oct 2015
58	03	28 Oct 2015	94	03	28 Oct 2015
59	03	28 Oct 2015	95	03	28 Oct 2015
60	03	28 Oct 2015	96	03	28 Oct 2015
61	03	28 Oct 2015	97	03	28 Oct 2015
62	03	28 Oct 2015	98	03	28 Oct 2015
63	03	28 Oct 2015	99	03	28 Oct 2015
64	03	28 Oct 2015	100	03	28 Oct 2015
65	03	28 Oct 2015	101	03	28 Oct 2015
66	03	28 Oct 2015	102	03	28 Oct 2015
67	03	28 Oct 2015	**Chapter 4 TOC**		
68	03	28 Oct 2015			

Page No.	Revision No.	Effective Date
i	03	28 Oct 2015
ii	03	28 Oct 2015
iii	03	28 Oct 2015
iv	03	28 Oct 2015
v	03	28 Oct 2015
vi	03	28 Oct 2015
Chapter 4		
1	0	01 Nov 2013
2	02	21 Jul 2015
3	02	21 Jul 2015
4	01	14 Nov 2014
5	0	01 Nov 2013
6	03	28 Oct 2015
7	03	28 Oct 2015
8	03	28 Oct 2015
9	0	01 Nov 2013
10	0	01 Nov 2013
11	01	14 Nov 2014
12	02	21 Jul 2015
13	02	21 Jul 2015
14	03	28 Oct 2015
15	03	28 Oct 2015
16	02	21 Jul 2015
17	02	21 Jul 2015
18	02	21 Jul 2015
19	02	21 Jul 2015
20	02	21 Jul 2015
21	02	21 Jul 2015
22	02	21 Jul 2015
23	02	21 Jul 2015
24	02	21 Jul 2015
25	02	21 Jul 2015
26	03	28 Oct 2015
27	03	28 Oct 2015
28	03	28 Oct 2015

Page No.	Revision No.	Effective Date
29	03	28 Oct 2015
30	03	28 Oct 2015
31	03	28 Oct 2015
32	03	28 Oct 2015
33	03	28 Oct 2015
34	03	28 Oct 2015
35	03	28 Oct 2015
36	03	28 Oct 2015
37	03	28 Oct 2015
38	03	28 Oct 2015
39	03	28 Oct 2015
40	03	28 Oct 2015
41	03	28 Oct 2015
42	03	28 Oct 2015
43	03	28 Oct 2015
44	03	28 Oct 2015
45	03	28 Oct 2015
46	03	28 Oct 2015
47	03	28 Oct 2015
48	03	28 Oct 2015
49	03	28 Oct 2015
50	03	28 Oct 2015
51	03	28 Oct 2015
52	03	28 Oct 2015
53	03	28 Oct 2015
54	03	28 Oct 2015
55	03	28 Oct 2015
56	03	28 Oct 2015
57	03	28 Oct 2015
58	03	28 Oct 2015
59	03	28 Oct 2015
60	03	28 Oct 2015
61	03	28 Oct 2015
62	03	28 Oct 2015
63	03	28 Oct 2015
64	03	28 Oct 2015

Page No.	Revision No.	Effective Date	Page No.	Revision No.	Effective Date
65	03	28 Oct 2015	101	03	28 Oct 2015
66	03	28 Oct 2015	102	03	28 Oct 2015
67	03	28 Oct 2015	103	03	28 Oct 2015
68	03	28 Oct 2015	104	03	28 Oct 2015
69	03	28 Oct 2015	105	03	28 Oct 2015
70	03	28 Oct 2015	106	03	28 Oct 2015
71	03	28 Oct 2015	107	03	28 Oct 2015
72	03	28 Oct 2015	108	03	28 Oct 2015
73	03	28 Oct 2015	109	03	28 Oct 2015
74	03	28 Oct 2015	110	03	28 Oct 2015
75	03	28 Oct 2015	111	03	28 Oct 2015
76	03	28 Oct 2015	112	03	28 Oct 2015
77	03	28 Oct 2015	113	03	28 Oct 2015
78	03	28 Oct 2015	114	03	28 Oct 2015
79	03	28 Oct 2015	115	03	28 Oct 2015
80	03	28 Oct 2015	116	03	28 Oct 2015
81	03	28 Oct 2015	117	03	28 Oct 2015
82	03	28 Oct 2015	118	03	28 Oct 2015
83	03	28 Oct 2015	119	03	28 Oct 2015
84	03	28 Oct 2015	120	03	28 Oct 2015
85	03	28 Oct 2015	121	03	28 Oct 2015
86	03	28 Oct 2015	122	03	28 Oct 2015
87	03	28 Oct 2015	123	03	28 Oct 2015
88	03	28 Oct 2015	124	03	28 Oct 2015
89	03	28 Oct 2015	125	03	28 Oct 2015
90	03	28 Oct 2015	126	03	28 Oct 2015
91	03	28 Oct 2015	127	03	28 Oct 2015
92	03	28 Oct 2015	128	03	28 Oct 2015
93	03	28 Oct 2015	129	03	28 Oct 2015
94	03	28 Oct 2015	130	03	28 Oct 2015
95	03	28 Oct 2015	131	03	28 Oct 2015
96	03	28 Oct 2015	132	03	28 Oct 2015
97	03	28 Oct 2015	133	03	28 Oct 2015
98	03	28 Oct 2015	134	03	28 Oct 2015
99	03	28 Oct 2015	135	03	28 Oct 2015
100	03	28 Oct 2015	136	03	28 Oct 2015

Page No.	Revision No.	Effective Date	Page No.	Revision No.	Effective Date
137	03	28 Oct 2015	173	03	28 Oct 2015
138	03	28 Oct 2015	174	03	28 Oct 2015
139	03	28 Oct 2015	175	03	28 Oct 2015
140	03	28 Oct 2015	176	03	28 Oct 2015
141	03	28 Oct 2015	177	03	28 Oct 2015
142	03	28 Oct 2015	178	03	28 Oct 2015
143	03	28 Oct 2015	179	03	28 Oct 2015
144	03	28 Oct 2015	180	03	28 Oct 2015
145	03	28 Oct 2015	181	03	28 Oct 2015
146	03	28 Oct 2015	182	03	28 Oct 2015
147	03	28 Oct 2015	183	03	28 Oct 2015
148	03	28 Oct 2015	184	03	28 Oct 2015
149	03	28 Oct 2015	185	03	28 Oct 2015
150	03	28 Oct 2015	186	03	28 Oct 2015
151	03	28 Oct 2015	187	03	28 Oct 2015
152	03	28 Oct 2015	188	03	28 Oct 2015
153	03	28 Oct 2015	189	03	28 Oct 2015
154	03	28 Oct 2015	190	03	28 Oct 2015
155	03	28 Oct 2015	191	03	28 Oct 2015
156	03	28 Oct 2015	192	03	28 Oct 2015
157	03	28 Oct 2015	193	03	28 Oct 2015
158	03	28 Oct 2015	194	03	28 Oct 2015
159	03	28 Oct 2015	195	03	28 Oct 2015
160	03	28 Oct 2015	196	03	28 Oct 2015
161	03	28 Oct 2015	197	03	28 Oct 2015
162	03	28 Oct 2015	198	03	28 Oct 2015
163	03	28 Oct 2015	199	03	28 Oct 2015
164	03	28 Oct 2015	200	03	28 Oct 2015
165	03	28 Oct 2015	201	03	28 Oct 2015
166	03	28 Oct 2015	202	03	28 Oct 2015
167	03	28 Oct 2015	203	03	28 Oct 2015
168	03	28 Oct 2015	204	03	28 Oct 2015
169	03	28 Oct 2015	205	03	28 Oct 2015
170	03	28 Oct 2015	206	03	28 Oct 2015
171	03	28 Oct 2015	207	03	28 Oct 2015
172	03	28 Oct 2015	208	03	28 Oct 2015

Page No.	Revision No.	Effective Date
209	03	28 Oct 2015
210	03	28 Oct 2015
211	03	28 Oct 2015
212	03	28 Oct 2015

Chapter 5 TOC

Page No.	Revision No.	Effective Date
i	03	28 Oct 2015
ii	01	14 Nov 2014

Chapter 5

Page No.	Revision No.	Effective Date
1	03	28 Oct 2015
2	01	14 Nov 2014
3	01	14 Nov 2014
4	01	14 Nov 2014
5	01	14 Nov 2014
6	01	14 Nov 2014
7	01	14 Nov 2014
8	01	14 Nov 2014
9	01	14 Nov 2014
10	01	14 Nov 2014
11	01	14 Nov 2014
12	01	14 Nov 2014
13	01	14 Nov 2014
14	01	14 Nov 2014
15	01	14 Nov 2014
16	01	14 Nov 2014
17	01	14 Nov 2014
18	01	14 Nov 2014
19	01	14 Nov 2014
20	01	14 Nov 2014
21	01	14 Nov 2014
22	01	14 Nov 2014
23	01	14 Nov 2014
24	01	14 Nov 2014
25	01	14 Nov 2014
26	01	14 Nov 2014
27	01	14 Nov 2014
28	01	14 Nov 2014
29	01	14 Nov 2014
30	01	14 Nov 2014
31	01	14 Nov 2014
32	01	14 Nov 2014
33	01	14 Nov 2014
34	01	14 Nov 2014
35	01	14 Nov 2014
36	01	14 Nov 2014
37	01	14 Nov 2014
38	01	14 Nov 2014
39	01	14 Nov 2014
40	01	14 Nov 2014
41	01	14 Nov 2014
42	01	14 Nov 2014

Chapter 6 TOC

Page No.	Revision No.	Effective Date
i	01	14 Nov 2014
ii	0	01 Nov 2013

Chapter 6

Page No.	Revision No.	Effective Date
1	01	14 Nov 2014
2	01	14 Nov 2014
3	01	14 Nov 2014
4	01	14 Nov 2014
5	01	14 Nov 2014
6	01	14 Nov 2014
7	01	14 Nov 2014
8	01	14 Nov 2014
9	01	14 Nov 2014
10	01	14 Nov 2014
11	01	14 Nov 2014
12	01	14 Nov 2014
13	01	14 Nov 2014
14	01	14 Nov 2014
15	01	14 Nov 2014
16	01	14 Nov 2014

Page No.	Revision No.	Effective Date
17	01	14 Nov 2014
18	03	28 Oct 2015
19	03	28 Oct 2015
20	03	28 Oct 2015
21	03	28 Oct 2015
22	03	28 Oct 2015
23	03	28 Oct 2015
24	03	28 Oct 2015
25	03	28 Oct 2015
26	03	28 Oct 2015
27	03	28 Oct 2015
28	03	28 Oct 2015
29	03	28 Oct 2015
30	03	28 Oct 2015
31	03	28 Oct 2015
32	03	28 Oct 2015
33	03	28 Oct 2015
34	03	28 Oct 2015
35	03	28 Oct 2015
36	03	28 Oct 2015
37	03	28 Oct 2015
38	03	28 Oct 2015
39	03	28 Oct 2015
40	03	28 Oct 2015
41	03	28 Oct 2015
42	03	28 Oct 2015
43	03	28 Oct 2015
44	03	28 Oct 2015
45	03	28 Oct 2015
46	03	28 Oct 2015
47	03	28 Oct 2015
48	03	28 Oct 2015
49	03	28 Oct 2015
50	03	28 Oct 2015
51	03	28 Oct 2015
52	03	28 Oct 2015
53	03	28 Oct 2015
54	01	14 Nov 2014
55	01	14 Nov 2014
56	01	14 Nov 2014

Chapter 7 TOC

Page No.	Revision No.	Effective Date
i	03	28 Oct 2015
ii	03	28 Oct 2015
iii	03	28 Oct 2015
iv	03	28 Oct 2015

Chapter 7

Page No.	Revision No.	Effective Date
1	01	14 Nov 2014
2	01	14 Nov 2014
3	01	14 Nov 2014
4	01	14 Nov 2014
5	01	14 Nov 2014
6	01	14 Nov 2014
7	01	14 Nov 2014
8	01	14 Nov 2014
9	01	14 Nov 2014
10	01	14 Nov 2014
11	01	14 Nov 2014
12	01	14 Nov 2014
13	01	14 Nov 2014
14	01	14 Nov 2014
15	01	14 Nov 2014
16	01	14 Nov 2014
17	01	14 Nov 2014
18	01	14 Nov 2014
19	03	28 Oct 2015
20	03	28 Oct 2015
21	01	14 Nov 2014
22	03	28 Oct 2015
23	01	14 Nov 2014
24	03	28 Oct 2015
25	01	14 Nov 2014

Page No.	Revision No.	Effective Date	Page No.	Revision No.	Effective Date
26	01	14 Nov 2014	62	01	14 Nov 2014
27	01	14 Nov 2014	63	01	14 Nov 2014
28	01	14 Nov 2014	64	01	14 Nov 2014
29	03	28 Oct 2015	65	01	14 Nov 2014
30	01	14 Nov 2014	66	01	14 Nov 2014
31	01	14 Nov 2014	67	01	14 Nov 2014
32	01	14 Nov 2014	68	01	14 Nov 2014
33	01	14 Nov 2014	69	01	14 Nov 2014
34	01	14 Nov 2014	70	01	14 Nov 2014
35	01	14 Nov 2014	71	01	14 Nov 2014
36	01	14 Nov 2014	72	01	14 Nov 2014
37	01	14 Nov 2014	73	01	14 Nov 2014
38	03	28 Oct 2015	74	01	14 Nov 2014
39	03	28 Oct 2015	75	03	28 Oct 2015
40	01	14 Nov 2014	76	03	28 Oct 2015
41	01	14 Nov 2014	77	01	14 Nov 2014
42	01	14 Nov 2014	78	01	14 Nov 2014
43	01	14 Nov 2014	79	01	14 Nov 2014
44	03	28 Oct 2015	80	01	14 Nov 2014
45	03	28 Oct 2015	81	01	14 Nov 2014
46	01	14 Nov 2014	82	03	28 Oct 2015
47	01	14 Nov 2014	83	01	14 Nov 2014
48	01	14 Nov 2014	84	03	28 Oct 2015
49	01	14 Nov 2014	85	01	14 Nov 2014
50	01	14 Nov 2014	86	03	28 Oct 2015
51	01	14 Nov 2014	87	03	28 Oct 2015
52	03	28 Oct 2015	88	01	14 Nov 2014
53	01	14 Nov 2014	89	01	14 Nov 2014
54	01	14 Nov 2014	90	01	14 Nov 2014
55	03	28 Oct 2015	91	03	28 Oct 2015
56	03	28 Oct 2015	92	01	14 Nov 2014
57	03	28 Oct 2015	93	01	14 Nov 2014
58	03	28 Oct 2015	94	01	14 Nov 2014
59	01	14 Nov 2014	95	01	14 Nov 2014
60	01	14 Nov 2014	96	01	14 Nov 2014
61	01	14 Nov 2014			

Page No.	Revision No.	Effective Date	Page No.	Revision No.	Effective Date
Chapter 8 TOC			26	0	01 Nov 2013
i	01	14 Nov 2014	27	03	28 Oct 2015
ii	01	14 Nov 2014	28	01	14 Nov 2014
iii	01	14 Nov 2014	29	0	01 Nov 2013
iv	01	14 Nov 2014	30	03	28 Oct 2015
v	01	14 Nov 2014	31	0	01 Nov 2013
vi	01	14 Nov 2014	32	03	28 Oct 2015
vii	01	14 Nov 2014	33	0	01 Nov 2013
viii	01	14 Nov 2014	34	03	28 Oct 2015
			35	03	28 Oct 2015
Chapter 8			36	0	01 Nov 2013
1	01	14 Nov 2014	37	0	01 Nov 2013
2	0	01 Nov 2013	38	0	01 Nov 2013
3	0	01 Nov 2013	39	0	01 Nov 2013
4	0	01 Nov 2013	40	0	01 Nov 2013
5	0	01 Nov 2013	41	0	01 Nov 2013
6	0	01 Nov 2013	42	0	01 Nov 2013
7	0	01 Nov 2013	43	01	14 Nov 2014
8	0	01 Nov 2013	44	01	14 Nov 2014
9	0	01 Nov 2013	45	0	01 Nov 2013
10	01	14 Nov 2014	46	0	01 Nov 2013
11	01	14 Nov 2014	47	0	01 Nov 2013
12	0	01 Nov 2013	48	01	14 Nov 2014
13	0	01 Nov 2013	49	01	14 Nov 2014
14	0	01 Nov 2013	50	01	14 Nov 2014
15	01	14 Nov 2014	51	03	28 Oct 2015
16	01	14 Nov 2014	52	0	01 Nov 2013
17	01	14 Nov 2014	53	0	01 Nov 2013
18	01	14 Nov 2014	54	01	14 Nov 2014
19	01	14 Nov 2014	55	0	01 Nov 2013
20	01	14 Nov 2014	56	01	14 Nov 2014
21	01	14 Nov 2014	57	0	01 Nov 2013
22	01	14 Nov 2014	58	01	14 Nov 2014
23	01	14 Nov 2014	59	01	14 Nov 2014
24	01	14 Nov 2014	60	01	14 Nov 2014
25	0	01 Nov 2013	61	0	01 Nov 2013

Page No.	Revision No.	Effective Date	Page No.	Revision No.	Effective Date
62	0	01 Nov 2013	98	01	14 Nov 2014
63	0	01 Nov 2013	99	01	14 Nov 2014
64	0	01 Nov 2013	100	01	14 Nov 2014
65	01	14 Nov 2014	101	01	14 Nov 2014
66	01	14 Nov 2014	102	01	14 Nov 2014
67	0	01 Nov 2013	103	01	14 Nov 2014
68	01	14 Nov 2014	104	01	14 Nov 2014
69	0	01 Nov 2013	105	01	14 Nov 2014
70	01	14 Nov 2014	106	01	14 Nov 2014
71	01	14 Nov 2014	107	01	14 Nov 2014
72	01	14 Nov 2014	108	01	14 Nov 2014
73	01	14 Nov 2014	109	01	14 Nov 2014
74	01	14 Nov 2014	110	01	14 Nov 2014
75	01	14 Nov 2014	111	01	14 Nov 2014
76	01	14 Nov 2014	112	01	14 Nov 2014
77	01	14 Nov 2014	113	01	14 Nov 2014
78	01	14 Nov 2014	114	01	14 Nov 2014
79	01	14 Nov 2014	115	01	14 Nov 2014
80	01	14 Nov 2014	116	01	14 Nov 2014
81	01	14 Nov 2014	117	01	14 Nov 2014
82	01	14 Nov 2014	118	01	14 Nov 2014
83	01	14 Nov 2014	119	01	14 Nov 2014
84	01	14 Nov 2014	120	3	28 Oct 2015
85	01	14 Nov 2014	121	01	14 Nov 2014
86	01	14 Nov 2014	122	01	14 Nov 2014
87	01	14 Nov 2014	123	01	14 Nov 2014
88	01	14 Nov 2014	124	01	14 Nov 2014
89	01	14 Nov 2014	125	01	14 Nov 2014
90	01	14 Nov 2014	126	01	14 Nov 2014
91	01	14 Nov 2014	127	01	14 Nov 2014
92	01	14 Nov 2014	128	01	14 Nov 2014
93	01	14 Nov 2014	129	01	14 Nov 2014
94	01	14 Nov 2014	130	01	14 Nov 2014
95	01	14 Nov 2014	131	01	14 Nov 2014
96	01	14 Nov 2014	132	01	14 Nov 2014
97	01	14 Nov 2014	133	01	14 Nov 2014

Page No.	Revision No.	Effective Date	Page No.	Revision No.	Effective Date
134	01	14 Nov 2014	170	01	14 Nov 2014
135	01	14 Nov 2014	171	01	14 Nov 2014
136	01	14 Nov 2014	172	3	28 Oct 2015
137	01	14 Nov 2014	173	01	14 Nov 2014
138	01	14 Nov 2014	174	01	14 Nov 2014
139	01	14 Nov 2014	175	3	28 Oct 2015
140	01	14 Nov 2014	176	01	14 Nov 2014
141	01	14 Nov 2014	177	01	14 Nov 2014
142	01	14 Nov 2014	178	01	14 Nov 2014
143	01	14 Nov 2014	179	01	14 Nov 2014
144	01	14 Nov 2014	180	01	14 Nov 2014
145	3	28 Oct 2015	181	01	14 Nov 2014
146	3	28 Oct 2015	182	01	14 Nov 2014
147	01	14 Nov 2014	183	01	14 Nov 2014
148	01	14 Nov 2014	184	01	14 Nov 2014
149	3	28 Oct 2015	185	01	14 Nov 2014
150	01	14 Nov 2014	186	01	14 Nov 2014
151	01	14 Nov 2014	187	01	14 Nov 2014
152	01	14 Nov 2014	188	01	14 Nov 2014
153	3	28 Oct 2015	189	01	14 Nov 2014
154	01	14 Nov 2014	190	01	14 Nov 2014
155	3	28 Oct 2015	191	01	14 Nov 2014
156	01	14 Nov 2014	192	01	14 Nov 2014
157	01	14 Nov 2014	193	01	14 Nov 2014
158	3	28 Oct 2015	194	01	14 Nov 2014
159	01	14 Nov 2014	195	01	14 Nov 2014
160	01	14 Nov 2014	196	01	14 Nov 2014
161	01	14 Nov 2014	197	01	14 Nov 2014
162	01	14 Nov 2014	198	01	14 Nov 2014
163	01	14 Nov 2014	199	01	14 Nov 2014
164	3	28 Oct 2015	200	01	14 Nov 2014
165	3	28 Oct 2015	201	01	14 Nov 2014
166	3	28 Oct 2015	202	01	14 Nov 2014
167	3	28 Oct 2015	203	01	14 Nov 2014
168	3	28 Oct 2015	204	01	14 Nov 2014
169	01	14 Nov 2014	205	01	14 Nov 2014

Page No.	Revision No.	Effective Date	Page No.	Revision No.	Effective Date
206	01	14 Nov 2014	242	01	14 Nov 2014
207	01	14 Nov 2014	243	01	14 Nov 2014
208	01	14 Nov 2014	244	01	14 Nov 2014
209	01	14 Nov 2014	245	01	14 Nov 2014
210	01	14 Nov 2014	246	01	14 Nov 2014
211	3	28 Oct 2015	247	01	14 Nov 2014
212	01	14 Nov 2014	248	01	14 Nov 2014
213	01	14 Nov 2014	249	03	28 Oct 2015
214	01	14 Nov 2014	250	01	14 Nov 2014
215	01	14 Nov 2014	251	01	14 Nov 2014
216	01	14 Nov 2014	252	01	14 Nov 2014
217	01	14 Nov 2014	253	01	14 Nov 2014
218	01	14 Nov 2014	254	01	14 Nov 2014
219	01	14 Nov 2014	255	01	14 Nov 2014
220	01	14 Nov 2014	256	01	14 Nov 2014
221	01	14 Nov 2014	257	01	14 Nov 2014
222	01	14 Nov 2014	258	01	14 Nov 2014
223	01	14 Nov 2014	259	01	14 Nov 2014
224	01	14 Nov 2014	260	01	14 Nov 2014
225	01	14 Nov 2014	261	01	14 Nov 2014
226	01	14 Nov 2014	262	01	14 Nov 2014
227	03	28 Oct 2015	263	01	14 Nov 2014
228	03	28 Oct 2015	264	01	14 Nov 2014
229	01	14 Nov 2014	265	01	14 Nov 2014
230	01	14 Nov 2014	266	01	14 Nov 2014
231	01	14 Nov 2014	267	01	14 Nov 2014
232	01	14 Nov 2014	268	01	14 Nov 2014
233	01	14 Nov 2014	269	01	14 Nov 2014
234	01	14 Nov 2014	270	01	14 Nov 2014
235	01	14 Nov 2014	271	01	14 Nov 2014
236	01	14 Nov 2014	272	01	14 Nov 2014
237	3	28 Oct 2015	273	01	14 Nov 2014
238	01	14 Nov 2014	274	01	14 Nov 2014
239	01	14 Nov 2014	275	01	14 Nov 2014
240	01	14 Nov 2014	276	01	14 Nov 2014
241	01	14 Nov 2014	277	01	14 Nov 2014

Page No.	Revision No.	Effective Date	Page No.	Revision No.	Effective Date
278	01	14 Nov 2014	314	01	14 Nov 2014
279	01	14 Nov 2014	315	3	28 Oct 2015
280	01	14 Nov 2014	316	3	28 Oct 2015
281	01	14 Nov 2014	317	01	14 Nov 2014
282	01	14 Nov 2014	318	3	28 Oct 2015
283	01	14 Nov 2014	319	3	28 Oct 2015
284	01	14 Nov 2014	320	3	28 Oct 2015
285	01	14 Nov 2014	321	01	14 Nov 2014
286	01	14 Nov 2014	322	01	14 Nov 2014
287	01	14 Nov 2014	323	3	28 Oct 2015
288	01	14 Nov 2014	324	01	14 Nov 2014
289	01	14 Nov 2014	325	3	28 Oct 2015
290	01	14 Nov 2014	326	01	14 Nov 2014
291	01	14 Nov 2014	327	01	14 Nov 2014
292	01	14 Nov 2014	328	01	14 Nov 2014
293	01	14 Nov 2014	329	01	14 Nov 2014
294	01	14 Nov 2014	330	01	14 Nov 2014
295	01	14 Nov 2014	331	01	14 Nov 2014
296	01	14 Nov 2014	332	01	14 Nov 2014
297	01	14 Nov 2014	333	01	14 Nov 2014
298	01	14 Nov 2014	334	01	14 Nov 2014
299	01	14 Nov 2014	335	01	14 Nov 2014
300	01	14 Nov 2014	336	01	14 Nov 2014
301	01	14 Nov 2014	337	01	14 Nov 2014
302	01	14 Nov 2014	338	01	14 Nov 2014
303	01	14 Nov 2014	339	01	14 Nov 2014
304	01	14 Nov 2014	340	01	14 Nov 2014
305	01	14 Nov 2014	341	01	14 Nov 2014
306	01	14 Nov 2014	342	01	14 Nov 2014
307	3	28 Oct 2015	343	01	14 Nov 2014
308	3	28 Oct 2015	344	01	14 Nov 2014
309	3	28 Oct 2015	345	01	14 Nov 2014
310	01	14 Nov 2014	346	01	14 Nov 2014
311	3	28 Oct 2015	347	01	14 Nov 2014
312	3	28 Oct 2015	348	01	14 Nov 2014
313	01	14 Nov 2014	349	01	14 Nov 2014

Page No.	Revision No.	Effective Date	Page No.	Revision No.	Effective Date

Approved / Accepted

MSP FSDO JOHN E LYONS Date: Digitally signed by JOHN E LYONS DN: c=US, o=U.S. Government, ou=AGL, ou=MSP FSDO, cn=JOHN E LYONS Date: 2015.11.05 10:14:34 -06'00'

Page No.	Revision No.	Effective Date	Page No.	Revision No.	Effective Date

Intentionally Left Blank

Master Table of Contents

Preface

Chapter 1: Reserved

Chapter 2: Limitations

Chapter 3: Normal Procedures

Chapter 4: Supplementary Procedures

Chapter 5: Environmental/Winter Operations

Chapter 6: Performance

Chapter 7: Standard Profiles & Maneuvers

Chapter 8: Emergency/Non-Normal

Preface

Reserved

Intentionally Left Blank

Chapter 8: Emergency/Non-Normal

Chapter 8: Emergency/Non-Normal

Preamble

14 CFR:	121.135(b)(12)	121.627(b)
CEME:	N	

General

This section of the manual supplies the procedures and guidance needed to successfully handle an emergency or non-normal situation.

It is structured so that all the required profiles, expanded procedures, and required checklists are contained in one section. This consolidates the information and makes it readily available during an emergency situation or during study or review.

In the event of a non-normal situation, the primary duty of the PF (pilot flying) is to fly the aircraft, confirm the identified malfunction, and call for the appropriate memory items or checklists. The PM (pilot monitoring) is to clearly identify the malfunction and complete all appropriate checklists as requested. So the PM is afforded an uninterrupted opportunity to complete all required checklists, responsibility for radio communications shall be assumed by the PF. Depending on the nature/severity of the malfunction the PM may be required to maintain or re-assume responsibility for the radios.

When a non-normal situation occurs remember that special care must be taken to correctly identify the problem. In the event an incorrect procedure is applied to a situation; not only will the original problem exist, but a secondary problem or operating restriction may be created.

In flight, prior to any movement of a thrust lever, flight control disconnect, any guarded switchlight, generator switch or fuel boost pump switchlight, the selection must be confirmed. Re-confirmation of a critical control will be made if the pilot removes their hand from the control prior to making the appropriate selection. On the ground, confirmation of critical controls is not required.

It is the captain's discretion as to who will fly during a non-normal situation based on the nature of the malfunction and experience of the first officer. If transfer of control is made, the transfer in roles and duties as PF and PM must also be clearly made.

To summarize, the thought process that should be followed in all emergency/ abnormal procedures is this:

FLY THE AIRCRAFT

Cancel the Aural
Identify the Malfunction
Read the Checklist

DO NOT HURRY

Malfunction Identification and Confirmation

All malfunctions must be dealt with in a specific, disciplined and coordinated manner. The tendency to over analyze or "overtalk" a situation should be avoided.

Normally, the PM will assess, cancel the aural and clearly identify the malfunction. The PF will then check and "confirm" the malfunction. Studies recommend being specific when making identification.

Example (Emergency Checklist card item):

PM:	"Left Engine Fire"
PF:	"Confirmed. Emergency Checklist - Left Engine Fire."

Example (Malfunction not on Emergency Checklist card):

PM:	"GEN 1 OFF"
PF:	"Confirmed. Non-Normal Checklist - GEN 1 OFF."

If the PM misidentifies a malfunction, the PF will clearly disagree and then state the problem as he/she sees it. The PM will then start the identification procedure over.

Example:

PM:	"GEN 2 OFF"
PF:	"Negative, GEN 1 OFF"
PM:	"GEN 1 OFF"
PF:	"Confirmed. Non-Normal Checklist - GEN1 OFF"

If agreement on the malfunction cannot be reached, then both pilots will discuss the situation to resolve the problem and facilitate a solution. It is important that the identification process not be rushed to avoid confusion and time loss.

NOTE:
A failure may have more than one associated EICAS message. If that is the case, system knowledge will have to be applied to identify the root cause and decide on which message is to be addressed first. The priority is: electrics then hydraulics then pneumatics then others.

Emergency/Non-Normal checklist are written based on worst case scenarios. When a checklist is complete, if the malfunction (message) no longer exists, any operating restrictions/limitations associated with the malfunction no longer apply and the aircraft may resume normal flight profiles.

Checklist Structure

Each checklist has a descriptive title. Items that are all capitalized are EICAS messages.

Example: **EFIS COMP MON**

Items that are procedural are presented with only the initial letter capitalized.

Example: **Aileron System Jammed**

Many of the notes address situations that may not apply. To avoid confusion when referencing notes, analyze the opening phrase of the sentence to see if it applies. If not, simply state "Not applicable" and continue to the next note.

Memory Items

Any actions or procedures that require immediate crew response are enclosed in a box and must be memorized. To further enhance crew coordination and CRM, procedures requiring memory items are expanded on in a PF/PM (Capt / FO) format.

NOTE:

Due to the urgency of some emergency situations associated with memory items, the normal method of identification and confirmation cannot be used. These are identified in the expanded procedures.

NOTE:

Boxed tables need not be memorized.

Emergency Items

Checklist items that require prompt crew action but are not required to be memorized are enclosed in a hatched box. These are available for quick reference on the red bordered Emergency Checklist Card in the flight deck.

Fault Resets

Some Emergency/Non-Normal checklist refer you to Fault Reset Procedures. When performing Fault Reset Procedures, the restrictions and limitations regarding their use apply. See Chapter 4 for additional information.

Some procedures require a different sequence of actions depending on the conditions observed (scenario). When this occurs, the statement "Choose a scenario" followed by choices in **bold** separated by light grey lines. Example:

AVIONIC FAN

Choose a scenario.

In flight:

Avionics Fan ....FLT ALTN

On the ground:

Do not takeoff.

Flap Selector ....Maintain current position

Black arrows indicate the path from one block of item(s) to the next and are generally associated with YES or NO answers to posed questions.

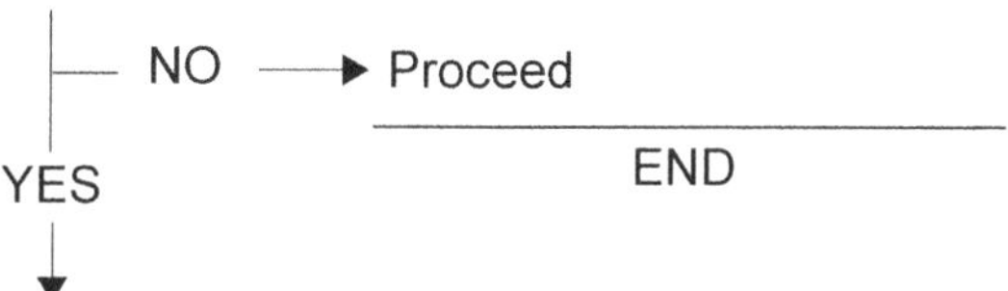

If a procedure has more than one set of actions, the completion of the procedure is indicated by a line with an END below. Unless directed to a note, the checklist is complete. Do not proceed any further.

If it is necessary to go to another procedure, the title and page number will be provided.

Each checklist is considered complete as indicated by a solid black line.

At the end of each page, if no solid black line is present, the procedure continues on to the next page.

Landing Distance Factors

Some non-normal checklists require increasing the Actual Landing Distance (ALD). These corrections are presented in table form as shown here:

Without two Thrust Reversers and/or Wet/Contaminated Runway Surface	With two Thrust Reversers and a Dry Runway Surface
1.30 (30%)	1.25 (25%)

Without two Thrust Reversers and/or Wet/Contaminated Runway Surface corrections are used when:

- Landing on a dry runway with one or both reverser(s) inoperative.
- Landing on a wet or contaminated runway.

With two Thrust Reversers and a Dry Runway Surface corrections are used ONLY when:

- Landing on a dry runway with BOTH thrust reversers operable.

In the case of multiple malfunctions, any applicable ALD factors must be multiplied together and the product used to correct the ALD as shown in the following example.

Example:

Aircraft is landing on a dry runway, single engine, with a GLD UNSAFE caution message displayed. The uncorrected, dry runway ALD is 3455 feet.

From the applicable non-normal checklists the Without Thrust Reversers (due to single engine condition) ALD factors are 1.30 and 1.35. To determine the ALD required, multiply the factors together then use the product to correct the ALD.

1.30 x 1.35 = 1.755

1.755 x 3455 ft. = 6064 ft.

Landings distances used in non-normal situations require the aircraft to cross the threshold at 50 feet at V_{REF}. Thrust should be reduced to idle at 50 feet. Touchdown should be firm with no float. Maximum braking consistent with runway length available should be used. If any thrust reverser is operative, maximum reverse should be used consistent with directional control.

Slope

Some landing weight penalties are slope dependent. To determine runway slope (%), divide the difference in height of the runway ends by the total length of the runway and multiply by 100.

V_{REF} Corrections

Many non-normal checklists require adjustments to V_{REF}. For multiple failures, only the highest applicable V_{REF} correction should be used. V_{REF} corrections are not additive.

Checklist Indexes

Index pages in this section have a complete list of all EICAS messages presented alphabetically and based on the following categories.Checklist Indexes

Index pages in this section have a complete list of all EICAS messages presented **alphabetically** and based on the following categories.

WARNING	W	Red
CAUTION	C	Amber
ADVISORY	A	Green
STATUS	S	White

For warning or caution messages, an associated page reference will direct you to the emergency / non-normal checklist. For advisory or status messages, a description of the message is given.

Example:

Message	Category	Reference
APU FIRE	W	7.1.29
APR INOP	C	7.1.14
HYD SOV 1 CLOSED	Hydraulic SOV 1 closed.	
L XFLOW ON	L cross flow on and auto fuel cross flow not activated.	

Index pages are also provided at the beginning of each system section. These reference checklist procedures not necessarily related to an EICAS message. For example control malfunctions, engine abnormalities, etc.

Example:

Flight Controls

Procedures

Circuit Breaker Reset Policy

Circuit breakers (CBs) are essentially heat-sensing protective devices. They protect the majority of electrical circuits on the airplane against electrical faults.

The CRJ-900 Electrical System is installed with trip-free-type circuit breakers. This means that if a trip condition exists, the breaker will open the faulty circuit, even if the circuit breaker is manually held in.

There is a latent danger in resetting a circuit breaker tripped by an unknown cause because the tripped condition is a signal that something may be wrong in the related electrical circuit.

Until it is positively determined what has caused a circuit breaker trip to occur, flight crews, maintenance personnel, or airplane ground servicing personnel usually have no way of knowing the consequences of resetting a tripped circuit breaker.

Resetting a circuit breaker tripped by a unknown cause should normally be a maintenance function conducted on the ground.

In Flight

A tripped circuit breaker should not be reset in-flight or cycled (i.e., opened then closed) unless doing so is consistent with explicit procedures specified in this chapter or unless, in the judgement of the PIC, that resetting or cycling the circuit breaker is necessary for the safe completion of the flight.

On Ground

A circuit breaker tripped by a unknown cause may be reset on the ground after maintenance has determined the cause of the trip and has determined that the circuit breaker may be safely reset.

A circuit breaker may be cycled (tripped or reset) as part of a troubleshooting procedure documented in this CFM, unless doing so is specifically prohibited for the conditions existing.

If the company MEL contains procedures that allow a tripped circuit breaker to be reset, then the same cautions identified above also apply.

Emergency Checklist Card

CRJ-900 Emergency Checklist

FIRE or SEVERE ENGINE DAMAGE

Engine Fire or Severe Engine Damage (In flight)

Affected thrust lever .... Confirm and IDLE

Affected thrust lever .... Confirm and SHUTOFF

Affected ENG FIRE PUSH switch ...... Confirm and select

Affected BOOST PUMP switch .... Confirm and OFF

If after 10 seconds the FIRE warning persists:

Affected engine BOTTLE switch......... Confirm and select
Push and hold until firex bottle light goes out.

If after 30 seconds the FIRE warning still persists:

Other engine BOTTLE switch............ Confirm and select
Push and hold until firex bottle light goes out.

Note: Verify that the hydraulic shut-off valve to the affected engine is closed. If open, confirm and select the affected HYD SOV switch to CLOSED.

Refer to CFM page 8-54 for In-Flight Engine Shutdown.

Engine Fire or Severe Engine Damage (On ground)

Note: Attempt to face the airplane into the wind.

Parking brake .... ON

Affected thrust lever............ Confirm and SHUTOFF

Affected ENG FIRE PUSH switch ...... Confirm and select

L and R BOOST PUMP switches .... Confirm and OFF

If after 10 seconds the FIRE warning persists:

Both engine BOTTLE switches .... Confirm and select
Push and hold until firex bottle light goes out.

Complete passenger evacuation as needed (found on back side of this card)

APU FIRE

APU FIRE PUSH switch............ Confirm and select

If after 5 seconds the FIRE warning persists:

APU BOTTLE switch............ Confirm and select

Refer to CFM page 8-91

ENGINE FAILURES

In-flight Engine Shutdown

Note: Covers all other in-flight engine failure or flameout scenarios not covered by the procedure above.

Affected thrust lever .... Confirm and IDLE

Affected thrust lever .... Confirm and SHUTOFF

Refer to CFM page 8-54

Double Engine Failure

CONT IGNITION switch .... ON

If engines continue to run down:

Thrust levers (both) .... Confirm and SHUTOFF

ADG manual deploy handle............ Pull

When ADG power is established:

STAB TRIM CH 2 .... Confirm engaged

Target Airspeed............ Establish
0.7 IMN above FL340
240 KIAS below FL340

Refer to CFM page 8-47

HYDRAULICS

HYD 1 HI TEMP

L HYD SOV switch .... Confirm and CLOSED

HYDRAULIC page............ Confirm L HYD SOV closed

System 1 temperature............ Monitor

Refer to CFM page 8-152

HYD 2 HI TEMP

R HYD SOV switch............ Confirm and CLOSED

HYDRAULIC page............ Confirm R HYD SOV closed

System 2 temperature............ Monitor

Refer to CFM page 8-157

Figure 8 - 1: Emergency Checklist Card (front)

CRJ-900 Emergency Checklist

OVERHEAT

MLG BAY OVHT

Airspeed .... Not more then 220 KIAS
Landing gear lever .... DN
Refer to CFM page 8-291

BRAKE OVHT

In flight:
Airspeed .... Not more than 220 KIAS
Landing gear lever .... DN
Refer to CFM page 8-284

On ground:
Refer to CFM page 8-284

MISCELLANEOUS

CABIN ALT or Emergency Descent

Oxygen masks .... Don, set to 100%
Crew communication .... Establish
PASS SIGNS switches (both) .... ON
Descent .... Initiate to 10,000 or lowest safe altitude
Thrust levers .... IDLE
Flight spoilers .... MAX
Refer to CFM page 8-237

Ditching or Forced Landing Imminent

L PACK and R PACK switches .... Select off
EMER DEPRESS switch .... Confirm and ON
Just before contact:
EMER DEPRESS switch .... Confirm and select off
Refer to CFM page 8-341

Passenger Evacuation

Captain:
Parking brake .... ON
Evacuation .... Command to FO
GND LIFT DUMPING switch .... MAN DISARM
Thrust levers .... SHUTOFF
Evacuation .... Initiate using PA system
APU, LH, and RH ENG,
FIRE PUSH switches .... Select
BATTERY MASTER switch .... OFF

First Officer (on evacuation command):
ATC .... Notify of emergency conditions and of intention to evacuate.

Note: If ditching, disregard EMER DEPRESS action.

EMER DEPRESS switch .... ON
EMER LTS switch .... ON with PA announcement

Both pilots:
Appropriate exits .... Open
Passenger evacuation .... Assist and direct passengers away from airplane
Airplane .... Abandon by any suitable exit

Figure 8 - 2: Emergency Checklist Card (back)

Intentionally Left Blank

EICAS Message Index

Intentionally Left Blank

EICAS Message Index

Warnings (Red)

Table 8 - 1: Warnings (Red)

Message	Type	Reference
AFCS MSG FAIL	W	8-307
ANTI-ICE DUCT	W	8-264
APU FIRE	W	8-91
APU OVERSPEED	W	8-92
APU OVERTEMP	W	8-93
BRAKE OVHT	W	8-284
CABIN ALT	W	8-237
CONFIG AILERON	W	8-321
CONFIG AP	W	8-321
CONFIG FLAPS	W	8-321
CONFIG RUDDER	W	8-321
CONFIG SPLRS	W	8-321
CONFIG STAB	W	8-321
DIFF PRESS	W	8-239
EMER PWR ONLY	W	8-99
ENGINE OVERSPD	W	8-53
GEAR DISAGREE	W	8-286
L BLEED DUCT	W	8-252
L COWL A/I DUCT	W	8-266
L ENG FIRE	W	8-45
L ENG OIL PRESS	W	8-52
L REV DEPLOYED	W	8-51
MLG BAY OVHT	W	8-291
NOSE DOOR OPEN	W	8-291
PARKING BRAKE	W	8-285
PASSENGER DOOR	W	8-331
R BLEED DUCT	W	8-252
R COWL A/I DUCT	W	8-266
R ENG FIRE	W	8-45

Table 8 - 1: Warnings (Red)		
Message	**Type**	**Reference**
R ENG OIL PRESS	W	8-52
R REV DEPLOYED	W	8-51
SMOKE AFT CARGO	W	8-138
SMOKE AFT LAV	W	8-137
SMOKE FWD CARGO	W	8-138
SMOKE FWD LAV	W	8-137
WING OVHT	W	8-263

Caution (Amber)

Table 8 - 2: Caution (Amber)

Message	Type	Reference
A/SKID INBD	C	8-292
A/SKID INBD and A/SKID OUTBD	C	8-294
A/SKID OUTBD	C	8-292
AC 1 AUTOXFER	C	8-107
AC 2 AUTOXFER	C	8-107
AC BUS 1	C	8-108
AC BUS 2	C	8-109
AC ESS BUS	C	8-110
AC SERV BUS	C	8-110
AFT CARGO DET	C	8-139
AFT CARGO DOOR	C	8-332
AFT CARGO OVHT	C	8-244
AFT CARGO SQB 1	C	8-140
AFT CARGO SQB 2	C	8-140
ALT LIMITER	C	8-253
ANTI-ICE DUCT	C	8-264
ANTI-ICE LOOP	C	8-271
AP PITCH TRIM	C	8-319
AP TRIM IS LWD	C	8-318
AP TRIM IS ND	C	8-318
AP TRIM IS NU	C	8-318
AP TRIM IS RWD	C	8-318
APR CMD SET	C	8-64
APU BATT OFF	C	8-111
APU BLEED ON	C	8-88
APU BTL LO	C	8-94
APU DOOR OPEN	C	8-90
APU ECU FAIL	C	8-89
APU FAULT	C	8-88
APU FIRE FAIL	C	8-94

Table 8 - 2: Caution (Amber)

Message	Type	Reference
APU GEN OFF	C	8-112
APU GEN OVLD	C	8-112
APU LCV CLSD	C	8-87
APU LCV OPEN	C	8-86
APU PUMP	C	8-88
APU SOV FAIL	C	8-89
APU SOV OPEN	C	8-89
APU SQB	C	8-139
AUTO PRESS	C	8-253
AV BAY DOOR	C	8-332
AVIONICS FAN	C	8-244
BATTERY BUS	C	8-113
BLEED MISCONFIG	C	8-246
BULK FUEL TEMP	C	8-188
CABIN ALT	C	8-254
CARGO BTL LO	C	8-138
CTR CARGO DOOR	C	8-332
DC BUS 1	C	8-115
DC BUS 2	C	8-116
DC EMER BUS	C	8-117
DC ESS BUS	C	8-119
DC SERV BUS	C	8-118
DISPLAY COOL	C	8-245
EFIS COMP INOP	C	8-312
EFIS COMP MON (IRS)	C	8-312
EFIS COMP MON (AHRS)	C	8-324
ELEVATOR SPLIT	C	8-208
ELT ON	C	8-323
EMER DEPRESS	C	8-254
EMER LTS OFF	C	8-323
ENG BTL 1 LO	C	8-79
ENG BTL 2 LO	C	8-79

Table 8 - 2: Caution (Amber)

Message	Type	Reference
FIRE SYS FAULT	C	8-139
FLAPS FAIL	C	8-214
FLAPS FAIL and SLATS FAIL	C	8-220
FLT SPLR DEPLOY	C	8-213
FUEL CH 1/2 FAIL	C	8-182
FUEL IMBALANCE	C	8-183
FWD CARGO DET	C	8-139
FWD CARGO DOOR	C	8-332
FWD CARGO SQB 1	C	8-139
FWD CARGO SQB 2	C	8-139
FWD SERVICE DOOR	C	8-332
GEN 1 OFF	C	8-120
GEN 2 OFF	C	8-120
GEN 1 OVLD	C	8-121
GEN 2 OVLD	C	8-121
GLD NOT ARMED	C	8-232
GLD UNSAFE	C	8-231
GND SPLR DEPLOY	C	8-229
HYD 1 HI TEMP	C	8-152
HYD 1 LO PRESS	C	8-143
HYD 1 LO PRESS and HYD 2 LO PRESS	C	8-167
HYD 1 LO PRESS and HYD 3 LO PRESS	C	8-174
HYD 2 HI TEMP	C	8-157
HYD 2 LO PRESS	C	8-145
HYD 2 LO PRESS and HYD 3 LO PRESS	C	8-170
HYD 3 HI TEMP	C	8-164
HYD 3 LO PRESS	C	8-148
HYD EDP 1A	C	8-165
HYD EDP 2A	C	8-165
HYD PUMP 1B	C	8-165

Table 8 - 2: Caution (Amber)

Message	Type	Reference
HYD PUMP 2B	C	8-166
HYD PUMP 3A	C	8-166
HYD PUMP 3B	C	8-167
HYD SOV 1 OPEN	C	8-167
HYD SOV 2 OPEN	C	8-167
IB BRAKE PRESS	C	8-295
IB FLT SPLRS	C	8-225
IB GND SPLRS	C	8-230
IB SPOILERON	C	8-227
ICE	C	8-263
ICE DETECT FAIL	C	8-274
IDG 1	C	8-121
IDG 2	C	8-121
ISOL FAIL	C	8-251
L AFT EMER DOOR	C	8-332
L AOA HEAT	C	8-277
L BLEED DUCT	C	8-252
L BLEED LOOP	C	8-250
L COWL A/I	C	8-268
L COWL A/I OPEN	C	8-269
L COWL LOOP	C	8-271
L ENG BLEED	C	8-249
L ENG DEGRADED	C	8-83
L ENG FLAMEOUT	C	8-67
L ENG SOV CLSD	C	8-184
L ENG SOV FAIL	C	8-184
L ENG SOV OPEN	C	8-184
L ENG SQB	C	8-139
L ENG SRG CLSD	C	8-78
L ENG SRG OPEN	C	8-79
L ENG TAT HEAT	C	8-85
L FADEC	C	8-68

Table 8 - 2: Caution (Amber)

Message	Type	Reference
L FADEC OVHT	C	8-69
L FIRE FAIL	C	8-48
L FUEL FILTER	C	8-188
L FUEL LO PRESS	C	8-185
L FUEL LO TEMP	C	8-188
L FUEL PUMP	C	8-186
L FWD EMER DOOR	C	8-332
L MAIN EJECTOR	C	8-186
L PACK AUTOFAIL	C	8-243
L PACK	C	8-243
L PACK TEMP	C	8-240
L PITOT HEAT	C	8-276
L REV INOP	C	8-63
L REV UNLOCKED	C	8-63
L REV UNSAFE	C	8-62
L SCAV EJECTOR	C	8-185
L START ABORT	C	8-67
L START VALVE	C	8-85
L STATIC HEAT	C	8-276
L STRT VLV OPEN (in flight)	C	8-74
L STRT VLV OPEN (on ground)	C	8-76
L THROTTLE	C	8-65
L WINDOW HEAT	C	8-278
L WING A/I	C	8-269
L WING A/I and R WING A/I	C	8-271
L WSHLD HEAT	C	8-278
L XFER SOV	C	8-187
LO FUEL	C	8-181
MACH TRIM	C	8-210
MAIN BATT OFF	C	8-122
MLG OVHT FAIL	C	8-297

Table 8 - 2: Caution (Amber)

Message	Type	Reference
NO STRTR CUTOUT (in flight)	C	8-70
NO STRTR CUTOUT (on ground)	C	8-72
OB BRAKE PRESS	C	8-295
OB FLT SPLRS	C	8-226
OB GND SPLRS	C	8-230
OB SPOILERON	C	8-228
OVBD COOL	C	8-244
OXY LO PRESS	C	8-259
PARK BRAKE SOV	C	8-297
PASS OXY ON	C	8-259
PAX DR LATCH	C	8-335
PAX DR OUT HNDL	C	8-333
PITCH FEEL	C	8-206
PROX SYS CHAN	C	8-301
PROX SYSTEM	C	8-299
R AFT EMER DOOR	C	8-332
R AOA HEAT	C	8-277
R BLEED DUCT	C	8-252
R BLEED LOOP	C	8-250
R COWL A/I	C	8-268
R COWL A/I OPEN	C	8-269
R COWL LOOP	C	8-271
R ENG BLEED	C	8-249
R ENG DEGRADED	C	8-83
R ENG FLAMEOUT	C	8-67
R ENG SOV CLSD	C	8-184
R ENG SOV FAIL	C	8-184
R ENG SOV OPEN	C	8-184
R ENG SQB	C	8-139
R ENG SRG CLSD	C	8-78
R ENG SRG OPEN	C	8-79

Table 8 - 2: Caution (Amber)

Message	Type	Reference
R ENG TAT HEAT	C	8-85
R FADEC	C	8-68
R FADEC OVHT	C	8-69
R FIRE FAIL	C	8-48
R FUEL FILTER	C	8-188
R FUEL LO PRESS	C	8-185
R FUEL LO TEMP	C	8-188
R FUEL PUMP	C	8-186
R FWD EMER DOOR	C	8-332
R MAIN EJECTOR	C	8-186
R PACK AUTOFAIL	C	8-243
R PACK	C	8-243
R PACK TEMP	C	8-240
R PITOT HEAT	C	8-276
R REV INOP	C	8-63
R REV UNLOCKED	C	8-63
R REV UNSAFE	C	8-62
R SCAV EJECTOR	C	8-185
R START ABORT	C	8-67
R START VALVE	C	8-85
R STATIC HEAT	C	8-276
R STRT VLV OPEN (in flight)	C	8-74
R STRT VLV OPEN (on ground)	C	8-76
R THROTTLE	C	8-65
R WINDOW HEAT	C	8-278
R WING A/I	C	8-269
R WSHLD HEAT	C	8-278
R XFER SOV	C	8-187
RUD LIMITER	C	8-207
SLATS FAIL	C	8-218

Table 8 - 2: Caution (Amber)

Message	Type	Reference
SPOILERONS ROLL	C	8-213
STAB TRIM	C	8-211
STAB TRIM LIMIT	C	8-212
STALL FAIL	C	8-205
STBY PITOT HEAT	C	8-277
STEERING INOP	C	8-297
TAT PROBE HEAT	C	8-278
WING A/I SNSR	C	8-272
WING XBLEED	C	8-272
WOW INPUT	C	8-302
WOW OUTPUT	C	8-304
XFLOW PUMP	C	8-188
YAW DAMPER	C	8-320

Advisory (Green)

Table 8 - 3: Advisory (Green)

ADS HEAT TEST OK	All L and R AOA, pitot, static, standby and TAT heater tests successful.
APU SOV CLSD	APU SOV confirmed closed and APU fire condition is detected.
CABIN ALT WARN HI	T/O or landing at high altitude > 8,000 ft.
COWL A/I ON	L and R cowl anti-ice selected on.
CPLT ROLL CMD	Copilot roll authority selected.
ENGS HI PWR SCHED	MCT applied to both engines if thrust levers in climb detent. APR applied to both engines if thrust levers in TOGA detent.
FDR EVENT	FDR EVENT switch pressed.
FIRE SYS OK	System in test, no failures detected.
FLAPS EMER	Emergency flap switch is in deploy position.

Table 8 - 3: Advisory (Green)

FLT SPLR DEPLOY	All the conditions that follow are present: - Any flight spoiler deployed (angle >3 degrees) or flight spoiler lever not at 0, - Radio altitude >300 ft., - Both engines N_1 <80%.
GLD MAN ARM	Ground lift dump manually armed.
GND SPLR DEPOLY	All the conditions that follow are present: - Any ground spoiler deployed, - Radio altitude ≤10 ft. Main landing gear weight-on-wheels.
GRAV XFLOW OPEN	Gravity cross flow valve confirmed open.
HYD SOV 1 CLSED	Hydraulic SOV 1 closed.
HYD SOV 2 CLSED	Hydraulic SOV 2 closed.
ICE	Ice detected by both ice detectors, wing and cowl anti-ice on, and operating normally.
L AUTO IGNITION	Left engine ignition commanded by FADEC.
L COWL A/I ON	Left cowl anti-ice selected on.
L ENG SOV CLSD	L engine fuel SOV confirmed closed and a fire condition is detected.
L FUEL PUMP ON	Left fuel boost pump operating.
L REV ARMED	Left thrust reverser armed
PARKING BRAKE ON	Parking brake set, aircraft on the ground, and one or both engines less than takeoff power, and IB brake pressure > 800 psi.
PLT ROLL CMD	Pilot roll authority selected.
R AUTO IGNITION	Right engine ignition commanded by FADEC.
R COWL A/I ON	Right cowl anti-ice selected on.
R ENG SOV CLSD	R engine fuel SOV confirmed closed and a fire condition is detected.
R FUEL PUMP ON	Right fuel boost pump operating.
R REV ARMED	Right thrust reverser armed.

Table 8 - 3: Advisory (Green)	
SPLR/STAB IN TEST	Spoiler and stabilizer systems are under test.
T/O CONFIG OK	All the conditions that follow are present: - Airplane is on ground, - Thrust reversers not deployed, - Autopilot not engaged, - Flaps and spoilers are in takeoff position, - Parking brake not set - Rudder trim < +/- 1 degree. - Aileron trim < +/- 1 degree. - Stabilizer trim is in the green band.
WING A/I ON	Wing anti-ice selected on and sufficient heat to both wings.
WING/COWL A/I ON	Wing and cowl anti-ice selected on and operating normally.

Status (White)

Table 8 - 4: Status (White)

Message	Description
A/SKID FAULT	Loss of redundancy of ASCU - loss of WOW input, spin down fail or loss of internal communication.
AC 1 AUTOXFER OFF	AC1 auto transfer selected off
AC 2 AUTOXFER OFF	AC2 auto transfer is selected off
AC ESS ALTN	AC essential bus-tie contactor is in alternate position.
ACARS CALL	ACARS has received a message.
ACARS MESSAGE	ACARS is receiving a message.
ACARS NOCOMM	ACARS is in NO COMM. Use Comm 2 for ARINC, or Company Radio for flight monitoring.
ADG AUTO FAIL	ADCU or up-lock solenoid failed or ADCU is unpowered with 3 seconds delay.
ADG FAIL	ADG GCU failure
AFT CARGO SOV	Cargo Air SOV failure.
APU ALT LIMIT	Surge control valve failed, APU operation limited to 17000' or less.
APU BATT CHGR	APU battery is overheating or not charging.
APU FAULT	Fault found in auxiliary power unit.
APU IN BITE	APU power fuel is selected and inlet door not positioned.
APU LCV OPEN	APU load control valve open.
APU SOV OPEN	APU fuel feed SOV confirmed open.
APU START	APU starter motor engaged.
AUTO PRESS 1 FAIL	Fault found on automatic pressurization system 1.
AUTO PRESS 2 FAIL	Fault found on automatic pressurization system 2.
AUTO PRS 1/2 FAIL	Indicates loss of active channel in manual mode.
AUTO XFLOW INHIB	Powered cross flow inhibited in automatic mode.
BLEED CLOSED	All Bleeds (L/R Engine and APU) CLOSED.

Table 8 - 4: Status (White)

Message	Description
BLEED MANUAL	Bleed System in MANUAL mode.
CABIN ALT WARN HI	T/O or landing at high altitude >8,000 ft.
CABIN PRESS MAN	Cabin pressurization is in manual control.
CABIN TEMP MAN	Cabin temperature in manual control.
CAS MISCOMP	Miscompare of Warning, Caution, Status, Advisory or Aural between DCUs for >20 seconds.
CKPT TEMP MAN	Cockpit temperature in manual control.
CONT IGNITION	Left and right ignition systems A and B are on.
COOL EXHAUST FAIL	Avionics exhaust fan failed, or low flow from cooling exhaust.
CPAM FAIL	CPCP indication system fail. Airplane altitude max FL300.
DC CROSS TIE CLSD	DC Cross-tie is closed
DC ESS TIE CLSD	DC essential tie is closed. **Note:** MEL item 24-31-1 allowing aircraft operation with a TRU failed, while that relief is in effect can result in the posting of this message.
DC MAIN TIE CLSD	DC main tie is closed. **Note:** MEL item 24-31-1 allowing aircraft operation with a TRU failed, while that relief is in effect can result in the posting of this message.
DCU 1 (2) AURAL INOP	Self detected internal aural fault or aural inoperative manually disabled. (See also page 8-323).
DCU1 (2) INOP	Self detected internal fault, or fault detected by other DCU's. (See also page 8-323).
DUCT MON FAULT	Loss of redundancy of leak detection system.
EMER LTS ON	Emergency lights are on, and power supply >4.5V.
ENG SYNC OFF	Engine synchronization manually deselected.
ESS TRU 1 FAIL	ESS TRU 1 <18V and AC ESS BUS powered or Ess Tie closed with ESS TRU 1 load <3.7A.

Table 8 - 4: Status (White)

Message	Description
ESS TRU 2 FAIL	Either of the conditions that follow are present: - ESS TRU 2 <18V and AC BUS 2 powered or ESS TRU 2 contactor not closed, - ESS TRU 2 contactor closed and AC ESS Bus powered, - ESS tie closed with ESS TRU 2 Load <3.7A.
ESS TRU 2 XFER	Essential TRU 2 transfer contactor closed.
FD 1 FAIL	Left flight director failure.
FD 2 FAIL	Right flight director failure.
FDR ACCEL FAIL	Triaxial accelerometer out of tolerance on ground with parking brake set and DC bus 1 on.
FDR FAIL	Difference between recorded data and data supplied by DCU.
FIRE SYS FAULT	Loss of redundancy in the FIREEX system.
FLAP FAULT	Fault found in flap system.
FLAPS HALFSPEED	Failure of one flap channel or when system operating on ADG power.
FLUTTER DAMPER	Left or right ailerons flutter damper failure (low fluid level).
FUEL CH 1 FAIL	Fuel quantity computer channel 1 failed.
FUEL CH 2 FAIL	Fuel quantity computer channel 2 failed.
FUEL QTY DEGRADED	Error in the attitude input to the fuel quantity gauging computer.
GLD MAN DISARM	Ground lift dump selected to MAN DISARM.
GPWS FAIL	Ground proximity warning system (GPWS) failed.
GRAV XFLOW FAIL	Gravity cross flow valve not confirmed to be opened or closed (failed) after 5 second time delay.
GS CANCEL	GPWS glideslope cancel mode selected.
HORN MUTED	Landing gear horn muted manually.
IAPS DEGRADED	IAPS bus failure, on ground, or both PSEUs invalid.

Table 8 - 4: Status (White)	
Message	**Description**
IAPS OVERTEMP	Over temperature condition found by any IAPS quadrant.
IB FLT SPLR FAULT	Loss of redundancy of roll assist control of inboard MFS. (On the ground, if message remains with IB (OB) SPOILERON caution message, consider fault reset attempt. See page 4-26.)
IB GND SPLR FAULT	Loss of redundancy found in inboard ground spoilers.
IB SPLRONS FAULT	Loss of redundancy found in inboard MFS spoilers for inboard roll assist.
ICE DET 1 FAIL	Ice detector 1 failed but ice detector 2 has not failed. (On the ground, if message remains, consider fault reset attempt. See page 4-26).
ICE DET 2 FAIL	Ice detector 2 failed but ice detector 1 has not failed. (On the ground, if message remains, consider fault reset attempt. See page 4-26.)
IDG 1 DISC	Integrated drive generator (IDG) 1 is disconnected (IDG 1 voltage <1.3 volts) and left engine N_2 >20%.
IDG 2 DISC	Integrated drive generator (IDG) 2 is disconnected (IDG 2 voltage <1.3 volts) and left engine N_2 >20%.
IRS 1 ATT	IRS 1 in attitude mode
IRS 2 ATT	IRS 2 in attitude mode
IRS 1 OVERTEMP	IRS 1 over temperature detected
IRS 2 OVERTEMP	IRS 2 over temperature detected
ISOL CLOSED	Cross bleed valve fully closed.
ISOL OPEN	Cross bleed valve fully open.
L AUTO XFLOW ON	Auto fuel cross flow activated and left cross flow on.
L COWL A/I DUCT	Left cowl anti-ice duct pressure < 3.12 psi or > 53.1 psi and the battery bus is powered.
L ENG BLEED CLSD	L Engine Bleed not selected LH PRSOV and HPSOV closed either in auto or manual.

Table 8 - 4: Status (White)

Message	Description
L ENG BLEED SNSR	Loss of redundancy of L Bleed system, Pack Inlet Pressure sensor fail.
L ENG SHUTDOWN	L engine has shutdown.
L ENG SQB	One left engine squib is failed or fired.
L ENGINE START	L engine start in progress.
L FADEC FAULT 1	Fault found in left FADEC that may affect flight performance. **Note**: If the messages L FADEC FAULT 1 and R FADEC FAULT 1 appear on EICAS simultaneously, return to gate. Maintenance check of the MDC is required.
L FADEC FAULT 2	Fault found in left FADEC, less serious than FAULT 1 status message.
L IGN A FAULT	Fault found in left engine ignition A.
L IGN B FAULT	Fault found in left engine ignition B.
L ITT EXCEEDED B	FADEC detected ITT B exceedance.
L ITT EXCEEDED B1	FADEC detected ITT B1 exceedance.
L ITT EXCEEDED C	FADEC detected ITT C exceedance.
L MLG FAULT	Fault in left main landing gear retraction actuator shuttle valve.
L OIL LEVEL LO	Left engine oil level low. (See page 8-36).
L PACK FAULT	Fault found in left pack.
L PACK OFF	Left pack selected off.
L RARV FAULT	Left Ram Air Regulating Valve failed.
L REV FAULT	Fault found in left thrust reverser.
L THROTTLE FAULT	L Throttle RVDT is faulty.
L VIB FAULT	L N_1 or N_2 speed sensor has failed.
L XFLOW ON	L cross flow on and auto fuel cross flow not activated.
MAIN BATT CHGR	Main battery overheating or is not charging.
MAN XFLOW	Manual cross flow selected.
MDC FAULT	MDC has detected an internal fault or has stopped sending valid data.

Table 8 - 4: Status (White)

Message	Description
MLG FAULT	PSEU detected a fault in the main landing gear shuttle valve.
NO SMOKING	No smoking sign is selected ON.
OB FLT SPLR FAULT	Loss of redundancy of roll assist control of outboard MFS. (On the ground, if message remains with IB (OB) SPOILERON caution message, consider fault reset attempt. See page 4-26).
OB GND SPLR FAULT	Loss of redundancy found in outboard ground spoilers.
OB SPLRONS FAULT	Loss of redundancy found in outboard MFS spoilers for outboard roll assist.
OUTFLOW VLV OPEN	Outflow valve is full open position.
OVBD COOL FAIL	Overboard avionics-cooling SOV failed closed and passenger door unlocked (open)(10 seconds time delay).
PITCH FEEL FAULT	Loss of redundancy of pitch feel system.
PROX SYS FAULT 1	Any one PSEU input or output related to a critical aircraft system failed.
PROX SYS FAULT 2	Ensure the main cabin door handle is in the fully OPEN position. Any one sensor or discrete input or output (non-critical) failed or unreasonable.
R AUTO XFLOW ON	Auto fuel cross flow activated and right cross flow on.
R COWL A/I DUCT	Right cowl anti-ice duct pressure < 3.12 psi or > 53.1 psi and the battery bus is powered.
R ENG BLEED CLSD	R Engine Bleed not selected RH PRSOV and HPSOV closed either in auto or manual.
R ENG BLEED SNSR	Loss of redundancy of R Bleed system, Pack Inlet Pressure sensor fail.
R ENG SHUTDOWN	R engine has shutdown.
R ENG SQB	One right engine squib is failed or fired.
R ENGINE START	R engine start in progress.

Table 8 - 4: Status (White)

Message	Description
R FADEC FAULT 1	Fault found in right FADEC that may affect flight performance. **Note**: If the messages L FADEC FAULT 1 and R FADEC FAULT 1 appear on EICAS simultaneously, return to gate. Maintenance check of the MDC is required.
R FADEC FAULT 2	Fault found in right FADEC, less serious than FAULT 1 message.
R IGN A FAULT	Fault found in right engine ignition A.
R IGN B FAULT	Fault found in right engine ignition B.
R ITT EXCEEDED B	FADEC detected ITT B exceedance.
R ITT EXCEEDED B1	FADEC detected ITT B1 exceedance.
R ITT EXCEEDED C	FADEC detected ITT C exceedance.
R MLG FAULT	Fault in right main landing gear retraction actuator shuttle valve.
R OIL LEVEL LO	Right engine oil level low. (See page 8-36).
R PACK FAULT	Fault found in right pack.
R PACK OFF	Right pack selected off.
R RARV FAULT	Right Ram Air Regulating Valve failed.
R REV FAULT	Fault found in right thrust reverser.
R THROTTLE FAULT	R throttle RVDT is faulty.
R VIB FAULT	R N_1 or N_2 speed sensor has failed.
R XFLOW ON	Right cross flow on and auto fuel cross flow not activated.
RAM AIR OPEN	Ram air SOV selected open. **Note**: MEL reliefs which require the aircraft to be operated unpressurized while that relief is in effect can result in the posting of this message.
RECIRC FAN FAULT	RECIRC FAN switch - OFF then ON. Either one or both of the recirculation fans have failed.
RECIRC FAN OFF	Recirculation fans selected off.

Table 8 - 4: Status (White)

Message	Description
RUD LIMIT FAULT	Loss of redundancy in rudder limiter. (On the ground, if message remains, consider fault reset attempt. See page 4-26). (On the ground, if message remains with SPLR/STAB FAULT status message, consider fault reset attempt. See page 4-26.)
SEAT BELTS	Seat belts signs selected on.
SLAT FAULT	L or R slat disconnect sensor has detected a mismatch.
SLATS HALFSPEED	Failure of one slat channel or when system operating on ADG power.
SPEED REFS INDEP	Pilot and copilot V-speed selection not synchronized or ADC crosstalk failed.
SPLR/STAB FAULT	One module in each SSCU has failed or the SPOST has not run for more than five consecutive times or problems identified during SPOST which impact the safety of the system.
SSCU 1 FAULT	SSCU 1 fault. (On the ground, if message remains, consider fault reset attempt. See page 4-26).
SSCU 2 FAULT	SSCU 2 fault. (On the ground, if message remains, consider fault reset attempt. See page 4-26.)
STAB CH1 INOP	STAB TRIM channel 1 not engaged with channel 2 engaged.
STAB CH2 INOP	STAB TRIM channel 2 not engaged with channel 1 engaged.
STAB FAULT	Loss of redundancy for the stabilizer trim channel.
STEERING DEGRADED	Possible intermittent loss of steering due to nose wheel bouncing.
TERRAIN FAIL	Enhanced GPWS terrain map failure.
TERRAIN NOT AVAIL	Enhanced GPWS terrain awareness display not available due to position accuracy.

Table 8 - 4: Status (White)

Message	Description
TERRAIN OFF	Enhanced GPWS terrain map manually deselected.
TRU 1 FAIL	Either of the conditions that follow are present: - TRU 1 <18V and AC BUS 1 powered - Main Tie contactor closed and TRU 1 load <3.7A.
TRU 2 FAIL	Either of the conditions that follow are present: - TRU 2 <18V and AC BUS 2 powered - Main Tie contactor closed and TRU 2 load <3.7A.
TRU FAN FAIL	Fan failed from any of 4 TRUs.(On the ground, if message remains, consider fault reset attempt. See page 4-26).
WINDSHEAR FAIL	Windshear detected subsystem (in GPWS) has failed or windshear guidance has failed.
WING A/I FAULT	Loss of redundancy on wing anti-ice system or loss of 1 or 2 outboard temperature sensors.
WING XBLEED OPEN	Wing cross bleed valve is in full open position.
YD 1 INOP	Yaw damper channel 1 is off or failed.
YD 2 INOP	Yaw damper channel 2 is off or failed.

L (R) OIL LEVEL LO" Status Message Explanation

1. The engine oil level can be checked at any time by pressing the MENU button once. The MENU page will be displayed on the EICAS Secondary Display (see below).

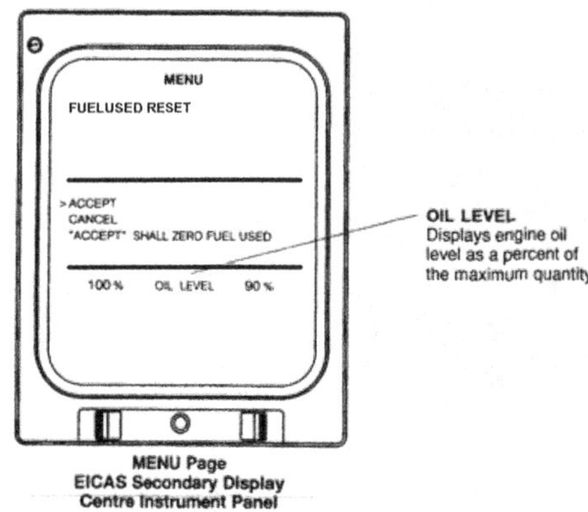

Figure 8 - 3: L (R) OIL LEVEL LO" STATUS MESSAGE

2. If the engine oil level falls below 80% (engines stopped) or below 57% (engines running), the associated L and R OIL LEVEL LO status message will be displayed.
3. Once the message is displayed, view the OIL LEVEL INDICATION and DURATION TABLE that follows. This chart provides information on the duration on flight available until oil servicing will be required by maintenance. As long as the engine oil level remains above 40% (engines stopped) or above 17% (engines running), the flight may be dispatched with the L or R OIL LEVEL LO message displayed.

Table 8 - 5: Oil Level Indication And Duration Table

Oil Level Indication And Duration Table		
Engine Oil Level Indication,%		
Engines Stopped	Engines Running	Duration Of Available Oil Until Next Service, Hours
100%	77%	36 hrs
80%	57%	26 hrs
50%	27%	10 hrs (1day)
40%	17%	5 hrs (1 flight)
<40% DO NOT DISPATCH	<17% DO NOT DISPATCH	
28% Do not operate, service the oil tank. (See note below)	15% Complete the flight, monitor oil temp and pressure.	***NOTE***: There is no EICAS OIL LEVEL indication if the oil quantity is less than 15%.
NOTE: The engine oil level check should be accomplished within 3 minutes to 2 hours after engine shutdown.		

4. The engine oil level check should normally be accomplished within 3 minutes to 2 hours after engine shutdown.
5. Once it has been determined that the oil level is above the DO NOT DISPATCH level, the aircraft may be dispatched. Upon first arrival at a maintenance base, if the oil level is less than 60% with the engines stopped or less than 37% with the engines running, an entry shall be made in the MM01 logbook.
6. An example of the L(R) ENG OIL PRESS and L(R) OIL LEVEL LO messages are provided on the following page.
7. Finally, when making the MM01 entry for the OIL LEVEL LO, always include the numbers shown on the EICAS menu page for both engines and the existing condition, either engines stopped or engines running.

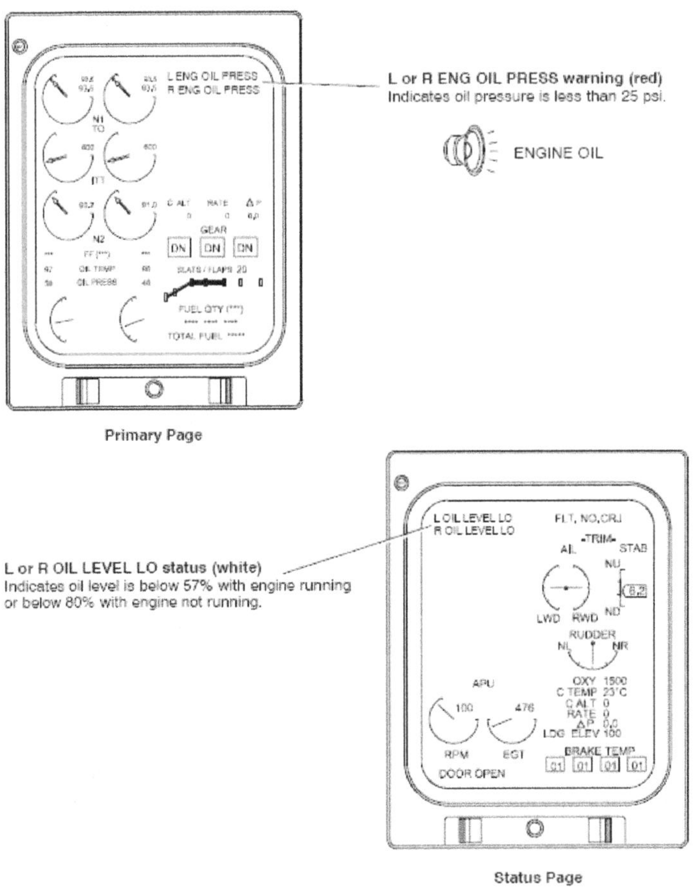

Figure 8 - 4: L (R) Oil Level Status

Non-EICAS Emergency/Abnormal Procedures

Table 8 - 6: Non-EICAS Emergency/Abnormal Procedures

Message	Reference
Abnormal Increase of Center Tank Quantity	8-191
AHRS Failure	8-323
Aileron or Rudder Trim Runaway	8-212
Aileron PCU Runaway	8-209
Aileron System Jammed	8-199
Air-Conditioning System Smoke	8-129
Air Data Computer (ADC) Failure	8-309
Altitude Alerting System Failure	8-317
Altitude Reporting Transponder Failure	8-317
APU Door Failure	8-90
Autopilot failure	8-307
Boost Pump Cycling	8-189
Cabin Smoke	8-135
Configuration Warning	8-321
Crew Escape Hatch Unsafe	8-332
Display Control Panel Failure	8-308
DISPLAY TEMP Flag	8-312
Ditching and Forced Landing	8-341
Ditching or Forced Landing Imminent	8-341
Double Engine Failure	8-47
During Landing - Excessive Asymmetry or Loss of Braking	8-285
EICAS Control Panel Failure	8-322
EICAS Primary Display Failure	8-321
EICAS Secondary Display Failure	8-321
Electrical System Smoke	8-132
Elevator System Jammed	8-201
Emergency Descent Procedure	8-237
Emergency Equipment Locations	8-349
Engine Failure During Approach	8-95
Engine Oscillations	8-83
Flight Director Guidance Failed	8-320
Flight Spoilers Lever Jammed (Spoilers Deployed)	8-213
Fuel Leak Procedure	8-193
Galley Smoke	8-134

Message	Reference
Gravity Crossfeed Procedure	8-190
High Oil Temperature	8-84
Hot Start	8-84
Ice Dispersal Procedure	8-267
In Flight Engine Shutdown	8-54
IRS Failure	8-316
Landing Gear Lever Jammed in the UP Position	8-288
Landing Gear Manual Extension	8-298
Landing Gear Up/Unsafe Landing Procedure	8-289
Loss of all AC Power (ADG not operative)	8-102
Loss of FAN VIB Indicator	8-85
Low Engine Oil Pressure Indication	8-52
Manual Bleed Procedure	8-247
Manual Pressurization Control Procedure	8-255
N1 Fan Vibration	8-80
N2 Core Vibration	8-82
Oxygen Pressure Readout Invalid (- - -)	8-259
Passenger Evacuation	8-348
Passenger Oxygen - Auto Deploy Failure	8-260
Pilot Incapacitation	8-260
Planned Ditching	8-342
Planned Forced Landing	8-345
Primary Flight Display Failure	8-307
Radio Altimeter (1 or 2) Failure	8-315
Radio Tuning Unit Failure	8-317
Rejected Takeoff	8-339
Rudder System Jammed	8-203
Severe Engine Damage (In flight)	8-45
Severe Engine Damage (On ground)	8-46
Single Engine Approach and Landing	8-62
SLATS/FLAPS Lever Jammed or Disconnected	8-224
Smoke or Fumes Removal Procedure	8-126
Smoke/Fire/Fumes Procedure	8-127
Stabilizer Trim Runaway	8-204
Suspected Leak into Center Tank	8-191
TCAS DISPLAY FAIL (Msg on MFD)	8-322

Message	Reference
TCAS FAIL (Msg on PFD/MFD)	8-322
TCAS RA FAIL (Msg on PFD)	8-322
Uncommanded Acceleration	8-53
Unpressurized Flight Procedure	8-256
Windshield/Window (Arcing, delaminated, shattered or cracked)	8-279

Intentionally Left Blank

Engines and APU Index

Engines and APU

L(R) ENG FIRE

Severe Engine Damage (In Flight)

Affected thrust lever .... Confirm and IDLE

Affected thrust lever ....Confirm and SHUT OFF

Affected ENG FIRE PUSH switch..... Confirm and select

Affected BOOST PUMP switch.......... Confirm and OFF

If after 10 seconds the fire warning persists:

Affected engine BOTTLE switch............................Select
Push and hold until
firex bottle light goes out

If after another 30 seconds the fire warning still persists:

Other engine BOTTLE switch................................Select
Push and hold until
firex bottle light goes out

NOTE:
Verify that the hydraulic shut-off valve to the affected engine is closed. If open, confirm and select the affected HYD SOV switch to CLOSED.

See "In-Flight Engine Shutdown" on page 8-54.

L(R) ENG FIRE

Severe Damage (On Ground)

NOTE:
Attempt to face the airplane into the wind.

PARKING BRAKE..On

Affected thrust lever .... Confirm and SHUT OFF

Affected ENG FIRE PUSH switch..... Confirm and select

FUEL, L and R BOOST PUMPS .... Confirm and off

If after 10 seconds the fire warning persists:

Both engine BOTTLE switches...............................Select
Push and hold until firex
bottle light goes out

Passenger Evacuation ....Consider

.See "Passenger Evacuation" on page 8- 348.

Double-Engine Failure

CONT IGNITION switch .... ON

If engines continue to run down ($N_2 < 40\%$):

Thrust levers (both) .... Confirm and SHUT OFF

ADG manual deploy handle .... Pull

When ADG power is established

STAB TRIM CH2 .... Confirm engaged

Target airspeed .... Establish
Above FL340 - 0.7 MACH
Below FL340 - 240 KIAS

Maintain airspeed until ready to restart engines.

Relight engines using APU bleed air whenever possible.

APU (if available, 37,000 feet and below) .... START

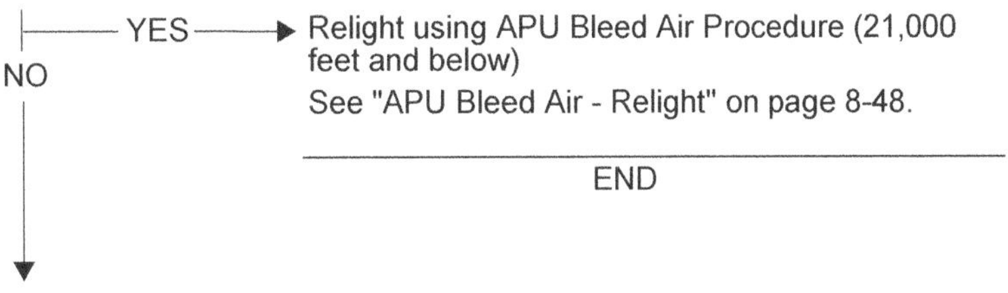

L (R) FIRE FAIL

Affected engine instruments .... Monitor

APU Bleed Air - Relight

From 21,000 feet and below:

Target airspeed..Re-establish

Airplane weight	Target Best Glide Speed
55,000 lb	170 KIAS
82,500 lb	210 KIAS

FUEL, L and R BOOST PUMPS ....Confirm ON

ANTI-ICE, WING and COWL...........................All OFF

BLEED SOURCE.. APU

ISOL..OPEN

BLEED VALVES...MANUAL

Attempt to start one engine at a time:

L or R ENG START ....Select and hold until N_2 is increasing

When N_2 is at least 20% and ITT is less than 90°C:

Applicable thrust lever.. ... IDLE

Engine indications.. Monitor\

CONTINUED ON NEXT PAGE

Did the engine relight and stabilize at flight idle?

YES →
- Thrust lever . . . . As required
- IGNITION, CONT . . . . .Off
- BLEED VALVES . . . . AUTO
- Re-establish normal power:

  Note: Confirm that at least one generator is operating.
- ADG manual deploy . . . . Stow
- ADG, PWR TXFR OVERRIDE . . . . Select
- See "Starter-Assisted APU Bleed Relight" on page 8-58 or "Single Engine Approach and Landing" on page 8-62 as required.

END

NO ↓

Affected engine thrust lever...... Confirm and SHUT OFF

Affected ENG STOP ....Select

Attempt relight on other engine. Return to "APU Bleed Air - Relight" on page 8-48.

If neither engine is restarted and altitude and time permit another relight:

Windmilling relight procedure .... Accomplish

See “Relight Using Windmilling Procedure” on page 8- 50.

If neither engine is restarted and altitude and time do not permit another relight:

Prepare for a forced landing or ditching. Notify the flight attendant.

See "Ditching and Forced Landing" on page 8-341.

Relight Using Windmilling Procedure

21,000 feet and below:

Attempt to start both engines at the same time.

Airspeed .... Not less than 250 KIAS

FUEL, L and R BOOST PUMPS .... Confirm ON

When engine rotation is established - minimum 7.2% N_2 and ITT is less than 90°C:

Thrust levers .... IDLE

NOTE:
An ENG FLAMEOUT caution message may momentarily be displayed.

Engine indications.. Monitor

NOTE:
N_2 acceleration should be positive and uninterrupted. Stable idle speed should be reached within 3 to 4 minutes.

Did at least one engine relight and stabilize at flight idle?

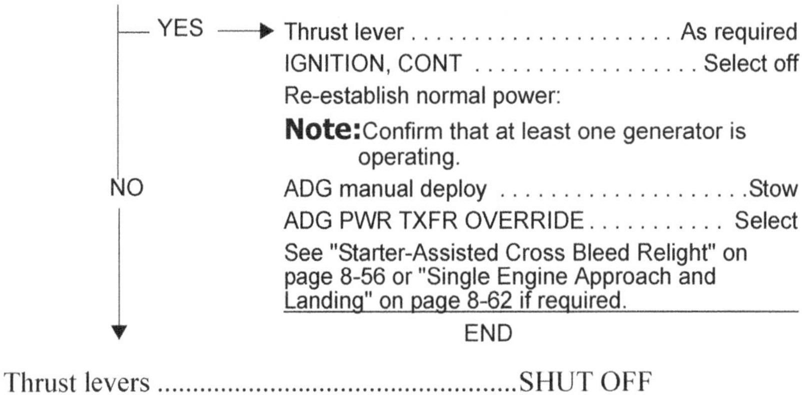

Thrust levers ....SHUT OFF

CONTINUED ON NEXT PAGE

Is another windmilling relight attempt still possible?

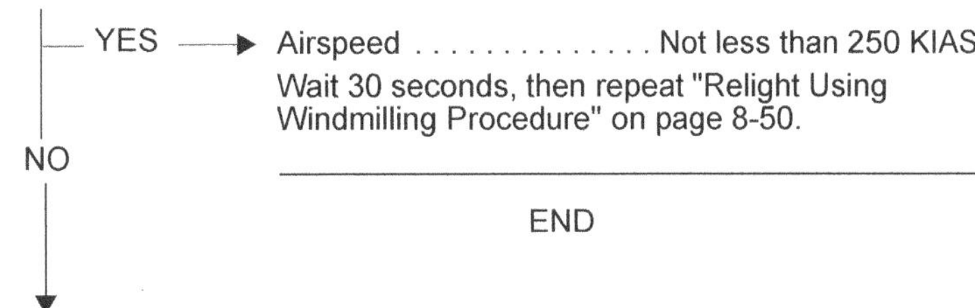

Prepare for a forced landing or ditching. Notify the flight attendant.

See "Ditching and Forced Landing" on page 8-341.

L (R) REV DEPLOYED

Affected thrust lever............................. Confirm & IDLE

Airspeed................................... Not more than 200 KIAS

Affected THRUST REVERSER switch.................... OFF

APU (if available, 37,000 feet and below) .... START

Engine Shutdown... Accomplish

See “In-Flight Engine Shutdown” on page 8- 54.

L (R) ENG OIL PRESS

OR

Low Engine Oil Pressure Indication

Affected engine oil pressure/quantity .... Check

Affected thrust lever .... Confirm and IDLE

Are any <u>two</u> of the following <u>three</u> indications displayed?

- L or R ENG OIL PRESS warning message;
- Affected engine oil pressure is below limits;
- Affected engine oil temperature is increasing or decreasing abnormally:

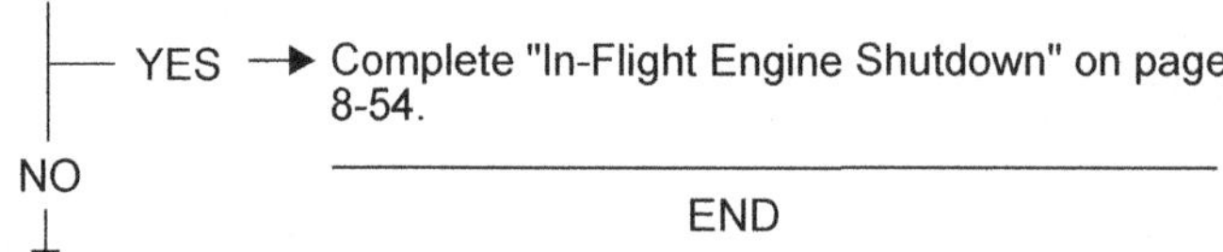

NOTE:
If the affected engine oil temperature is between 156°C and 163°C (amber range) or 164°C and higher (red range) accomplish the High Oil Temperature abnormal procedure. See "High Oil Temperature" on page 8- 84.

Affected thrust lever .... As required

Engine Indications .... Monitor

ENGINE OVERSPD

OR

Uncommanded Acceleration

Affected thrust lever..............................Confirm & IDLE

Engine indications .... Monitor

Does the engine respond?

YES → Affected thrust lever. . . . . .Adjust as required

END

NO ↓

See "In-Flight Engine Shutdown" on page 8-54.

NOTE:
Do not attempt to relight the engine.

In-Flight Engine Shutdown

Accomplish an engine shutdown only when flight conditions permit.

Affected thrust lever............................Confirm and IDLE

Affected thrust lever..................Confirm and SHUT OFF

Affected HYDRAULIC pump:

If left engine shut down.......................HYDRAULIC 1 switch ON

If right engine shut down ....HYDRAULIC 2 switch ON

Affected BOOST PUMP switch .... Confirm and off

XFLOW AUTO OVERRIDE switch .... Off (out)

Fuel quantity/balance.. Check

NOTE:
Powered crossflow may not be able to correct fuel imbalance during single engine operations.

WING A/I CROSS BLEED selector ....Select non-affected side

ANTI-ICE, LH or RH COWL ....Affected side OFF

NOTE:
Leave icing conditions to prevent ice accumulation on the engine cowl with the inoperative anti-icing system.

Continued on next page

Is enroute terrain clearance a consideration?

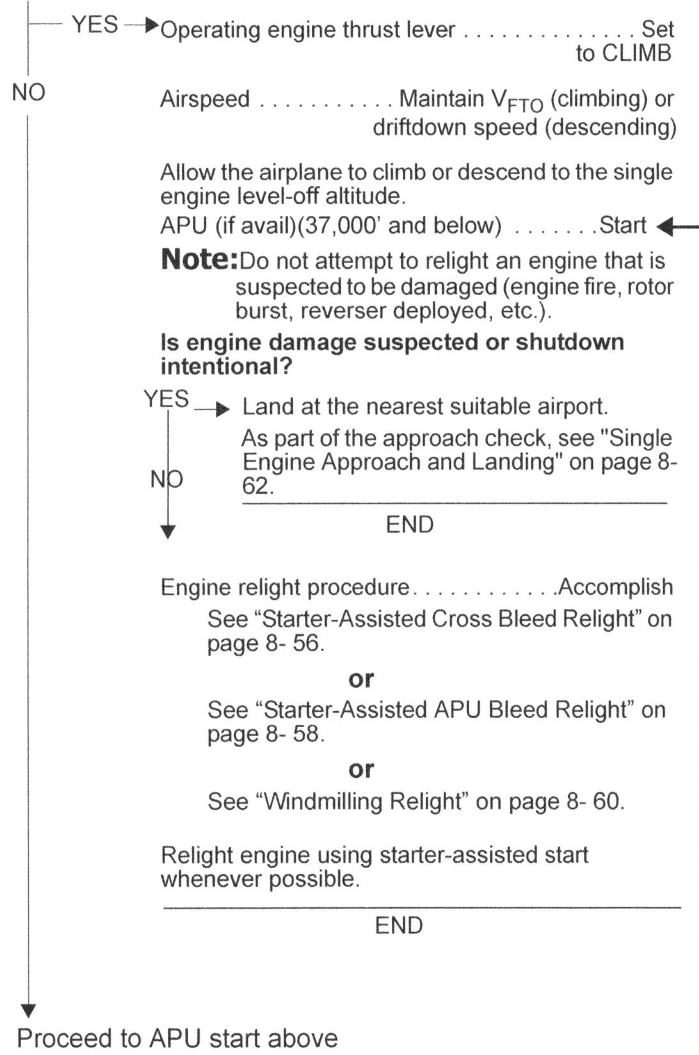

YES → Operating engine thrust lever . . . . Set to CLIMB

Airspeed . . . . Maintain V_{FTO} (climbing) or driftdown speed (descending)

Allow the airplane to climb or descend to the single engine level-off altitude.

APU (if avail)(37,000' and below) . . . . . . .Start

Note: Do not attempt to relight an engine that is suspected to be damaged (engine fire, rotor burst, reverser deployed, etc.).

Is engine damage suspected or shutdown intentional?

YES → Land at the nearest suitable airport.

As part of the approach check, see "Single Engine Approach and Landing" on page 8-62.

END

NO ↓

Engine relight procedure. . . . . .Accomplish

See "Starter-Assisted Cross Bleed Relight" on page 8- 56.

or

See "Starter-Assisted APU Bleed Relight" on page 8- 58.

or

See "Windmilling Relight" on page 8- 60.

Relight engine using starter-assisted start whenever possible.

END

NO ↓

Proceed to APU start above

Starter-Assisted Cross Bleed Relight

Altitude .... 21,000 feet and below

FUEL, L and R BOOST PUMPS .... Confirm ON

BLEED SOURCE................................ Operative engine

ISOL.. OPEN

BLEED VALVES.. MANUAL

Bleed air pressure............................. Not less than 40 PSI

When ready to start: ◄

IGNITION, CONT.. ON

Affected ENG START.. START
and hold until N_2 is increasing

When N_2 is at least 20% and ITT is less than 90°C:

Affected thrust lever .... IDLE

Engine indications.. Monitor

Does engine relight and stabilizes at IDLE:

— YES → Thrust lever . . . . As required
IGNITION, CONT . . . . Select OFF

Note: A BLEED MISCONFIG caution message will be displayed after engine start. The message will go out when the BLEED VALVES is selected to AUTO.

BLEED VALVES . . . . AUTO
WING A/I CROSS BLEED . . . . NORMAL
ANTI-ICE, WING and COWL . . . . As req'd
Affected HYDRAULIC pump . . . . AUTO

END

NO ↓

Affected thrust lever .... Confirm and SHUT OFF

Affected engine stop .... STOP

Wait 30 seconds, then repeat relight procedure if desired (return to "When ready to start")

CONTINUED ON NEXT PAGE

Engine relight still unsuccessful:

IGNITION, CONT .... Select OFF

Land at nearest suitable airport.

Single Engine Approach
and Landing procedure .... Accomplish

See "Single Engine Approach and Landing" on page 8- 62.

Starter-Assisted APU Bleed Relight

Altitude .... 21,000 feet or below

FUEL, L and R BOOST PUMPS .... Confirm ON

ANTI-ICE, WING and COWL........................... All OFF

BLEED SOURCE.. APU

ISOL... OPEN

BLEED VALVES.. MANUAL

When ready to start:

IGNITION, CONT... ON

Affected ENG START....................................... START
and hold until N_2 is increasing

When N_2 is at least 20% and ITT is less than 90°C:

Affected thrust lever .... IDLE

Engine indications... Monitor

Does engine relight and stabilize at idle?

YES →

Thrust lever. . . . . As required
IGNITION, CONT . . . . Select OFF
BLEED VALVES . . . . AUTO
WING A/I CROSS BLEED . . . . NORMAL
ANTI-ICE, WING and COWL . . . . As required
Affected HYDRAULIC pump . . . . AUTO
APU. . . . . As required

END

NO ↓

Affected thrust lever .... Confirm and SHUT OFF

Affected ENG STOP ....STOP

Wait 30 seconds, then repeat relight procedure, if desired. (→ return to "When ready to start:")

Engine relight still unsuccessful:

IGNITION, CONT... Select OFF

BLEED VALVES.. AUTO

CONTINUED ON NEXT PAGE

WING A/I CROSS BLEED....................................Select non-affected side

WING and non-affected COWL ANTI-ICE ...... As req'd

Land at the nearest suitable airport.

Single Engine Approach and

Landing Procedure... Accomplish

See "Single Engine Approach and Landing" on page 8-62.

Windmilling Relight

Altitude .... 21,000 feet or below

FUEL, L and R BOOST PUMPS .... Confirm ON

When ready to start:

IGNITION, CONT...ON

Airspeed..................................... Not less than 250 KIAS

NOTE:
Windmill airstart efficiency is enhanced by attaining the highest practical airspeed and N_2 within the relight envelope.

NOTE:
Maintain airspeed throughout light-off until the engine start is complete (stable idle). Monitor engine parameters carefully.

When N_2 is at least 7.2% and ITT is less than 90°C:

Affected thrust lever .... IDLE

NOTE:
An ENG FLAMEOUT caution message may momentarily be displayed.

Engine indications... Monitor

NOTE:
N_2 acceleration should be positive and uninterrupted. Stable idle speed should be reached within 3-4 minutes.

CONTINUED ON NEXT PAGE

Does engine relight and stabilize at idle?

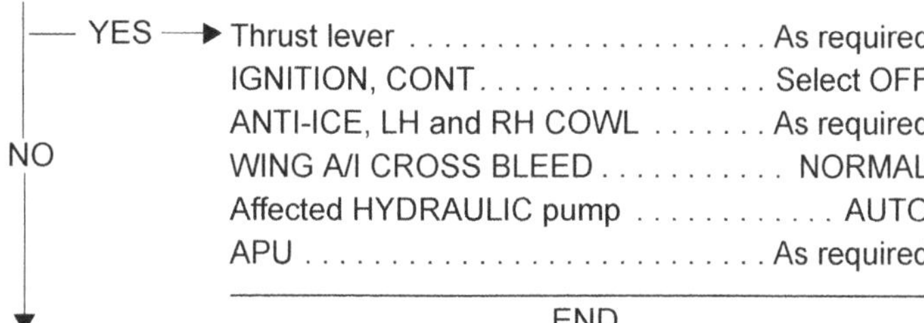

Affected thrust lever .... Confirm and SHUT OFF

Airspeed .... Not less than 250 KIAS

Wait 30 seconds, then repeat relight procedure, if desired.

Engine relight still unsuccessful:

IGNITION, CONT .... Select off

Land at the nearest suitable airport.

See "Single Engine Approach and Landing" on page 8-62

Single Engine Approach and Landing

GRND PROX, FLAP..OVRD

Approach and landing flaps.................................Use 20°

Final Approach speed .... Not less than V_{REF} (Flaps 45°)+12 KIAS

Actual landing distance ....Increase

Without two Thrust Reversers and/or Wet/Contaminated Runway Surface	With two Thrust Reversers and a Dry Runway Surface
1.30 (30%)	1.25 (25%)

CAUTION:
If required, use remaining thrust reverser carefully upon landing

L (R) REV UNSAFE

Affected THRUST REVERSER ....OFF

CAUTION:
If required, use remaining thrust reverser carefully upon landing.

L (R) REV INOP

Affected THRUST REVERSER .... Confirm ARMED

Does the REV INOP caution message persist?

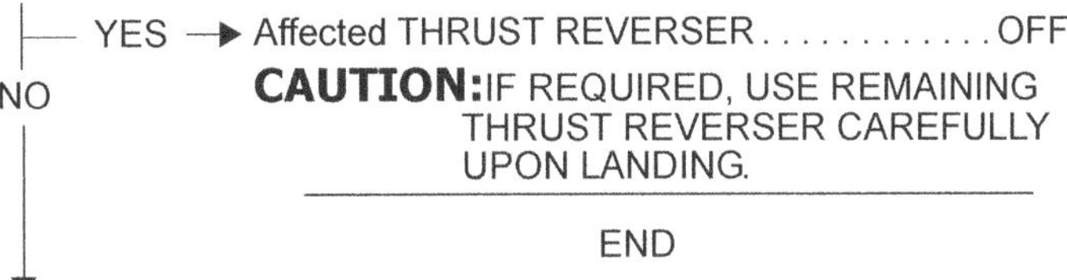

No further action required

L (R) REV UNLOCKED

Affected thrust lever .... Confirm and IDLE

Airspeed .... Maximum 200 KIAS

Affected THRUST REVERSER .... OFF

Prior to landing:

GRND PROX FLAP... OVRD

Landing flaps .... Use 20°

Approach speed .... $V_{REF\ (Flaps\ 45°)}$+12 KIAS

Actual landing distance .... Increase

Without two Thrust Reversers and/or Wet/Contaminated Runway Surface	With two Thrust Reversers and a Dry Runway Surface
1.35 (35%)	1.25 (25%)

CAUTION:
If required, use remaining thrust reverser carefully upon landing.

APR CMD SET

On the ground:

Do not takeoff.

In flight:

ENG SYNC...OFF

Thrust levers ....Adjust

NOTE:
Thrust levers may be split.

Engine Performance... Monitor

NOTE:
Selecting TOGA detent may set APR thrust.

NOTE:
Selecting CLIMB detent may set maximum continuous thrust.

L (R) THROTTLE

NOTE:
There will be no response to thrust lever movement.

NOTE:
FADEC will maintain the affected engine operating at a thrust level equal to the last valid thrust setting.

NOTE:
When either the landing gear is selected DN or the flaps are selected to greater than FLAPS 20, the FADEC will set the affected engine thrust to approach IDLE until touchdown and then to normal ground IDLE.

NOTE:
The thrust reverser on the affected side is inoperative.

NOTE:
Selecting the thrust lever to SHUT OFF will cause the engine to shut down.

Airplane performance .... Evaluate for continued engine operation at the current thrust setting.

Affected THRUST REVERSER .... Leave OFF

CONTINUED ON NEXT PAGE

Prior to approach:

Is the affected engine thrust above IDLE?

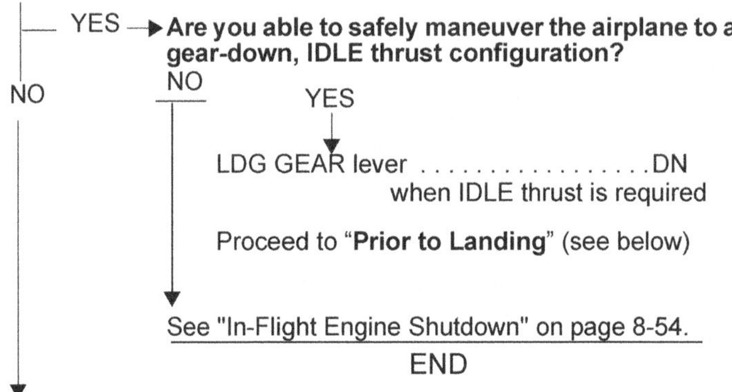

Affected engine thrust is at IDLE:

Prior to landing:

GRND PROX FLAP... OVRD

Landing flaps ....Use 20°

Approach speed........................ $V_{REF\ (Flaps\ 45°)}$+12 KIAS

Actual landing distance .... Increase

Without two Thrust Reversers and/or Wet/Contaminated Runway Surface	With two Thrust Reversers and a Dry Runway Surface
1.35 (35%)	1.25 (25%)

CAUTION:
If required, use remaining thrust reverser carefully upon landing

L (R) ENG FLAMEOUT

Affected Thrust lever .... Confirm and IDLE

Engine instruments .... Monitor auto-relight

NOTE:
Affected engine's ITT may drop; N_2 may drop below idle.

NOTE:
Affected engine may shutdown.

Does ENG FLAMEOUT caution message persist or N_2 drop below 40%?

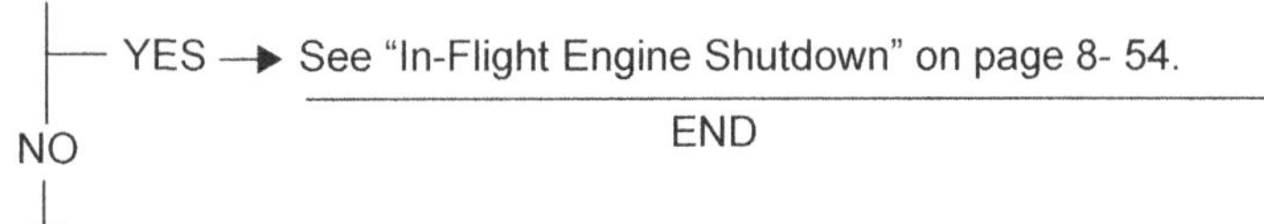

No further action required.

L (R) START ABORT

Affected thrust lever .... Confirm and SHUT OFF

Dry motor .... Until ITT < 120° or starter limit, whichever comes first

Affected ENG STOP .... STOP

Affected engine start .... Repeat

L (R) FADEC

Affected thrust lever .... Confirm and IDLE

Engine indications... Monitor

NOTE:
Engine may operate normally but without overspeed protection.

NOTE:
Engine RPM may reduce to idle setting.

NOTE:
Engine may shutdown.

Are all engine indications normal?

YES → Prior to landing:

GRND PROX FLAP .... OVRD

Affected THRUST REVERSER............ OFF

Landing flaps...................... Use 20°

Approach speed .... $V_{REF\ (Flaps\ 45°)}$+12 KIAS

Actual landing distance .... Increase

Without Thrust Reversers and/or Wet/Contaminated Runway Surface	With one thrust reverser and a Dry Runway Surface
1.35 (35%)	1.25 (25%)

CAUTION: IF REQUIRED, USE REMAINING THRUST REVERSER CAREFULLY UPON LANDING.

NO ↓

END

See "In-Flight Engine Shutdown" on page 8- 54.

L (R) FADEC OVHT

NOTE:
This message indicates possible erroneous information or loss of an engine.

Affected thrust lever ....Confirm and retard until message out

Leave icing conditions.

NOTE:
Icing conditions exist in-flight at a TAT of 10°C (50°F) or below, and visible moisture in any form is encountered (such as clouds, rain, snow, sleet or ice crystals), except when the SAT is -40°C (-40°F) or below.

When clear of icing conditions:

ANTI-ICE, LH or RH COWL ....Affected side OFF

Does the FADEC OVHT caution message persist?

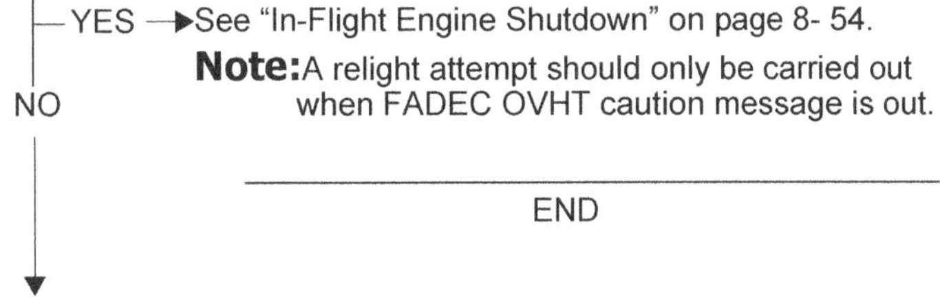

Engine Instruments .... Monitor

NO STRTR CUTOUT (In flight)

Choose a scenario

APU or CROSS BLEED START:

Affected ENG STOP ....STOP

Does the NO STRTR CUTOUT / STRT VLV OPEN caution message persist?

YES → BLEED SOURCE......................... Select alternate source

ISOL.. CLSD

BLEED VALVES...MANUAL

Does the NO STRTR CUTOUT / STRT VLV OPEN caution message still persist?

NO

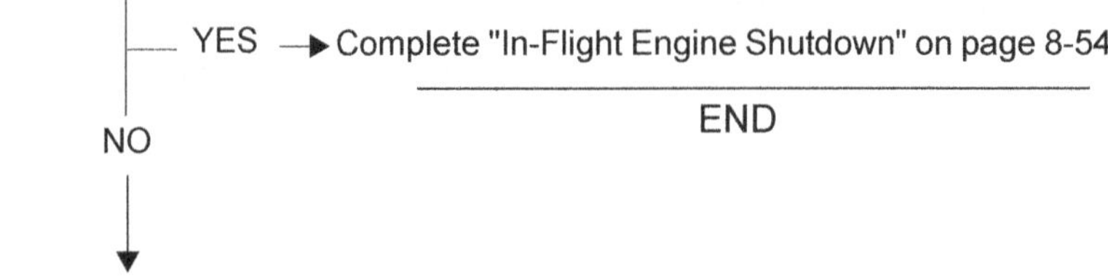

Inoperative PACK...OFF

NOTE:
Airplane altitude not above 25,000 feet during single pack operations with the engine as the bleed source.

WING A/I CROSS BLEED...... Select source engine side

ANTI-ICE, LH or RH COWL (Non-source engine side). ..... OFF

Leave icing conditions to prevent ice accumulation on inoperative cowl.

CONTINUED ON NEXT PAGE

NOTE:
Icing conditions exist in flight at a TAT of 10°C (50°F) or below and visible moisture in any form is encountered (such as clouds, rain, snow, sleet or ice crystals), except when the SAT is -40°C (-40°F) or below.

END

NO

Normal operations ....Continue

NO START BEING ATTEMPTED:

Engine Indications .... Monitor

Land at the nearest suitable airport.

Are significant abnormal indications occurring?

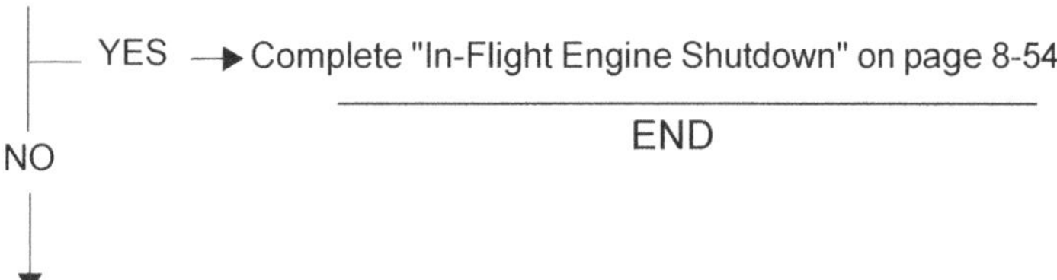

No further action required.

NO STRTR CUTOUT (On the Ground)

Choose a scenario

APU or CROSS BLEED START:

Affected ENG STOP ....STOP

Does the NO STRTR CUTOUT caution message persist?

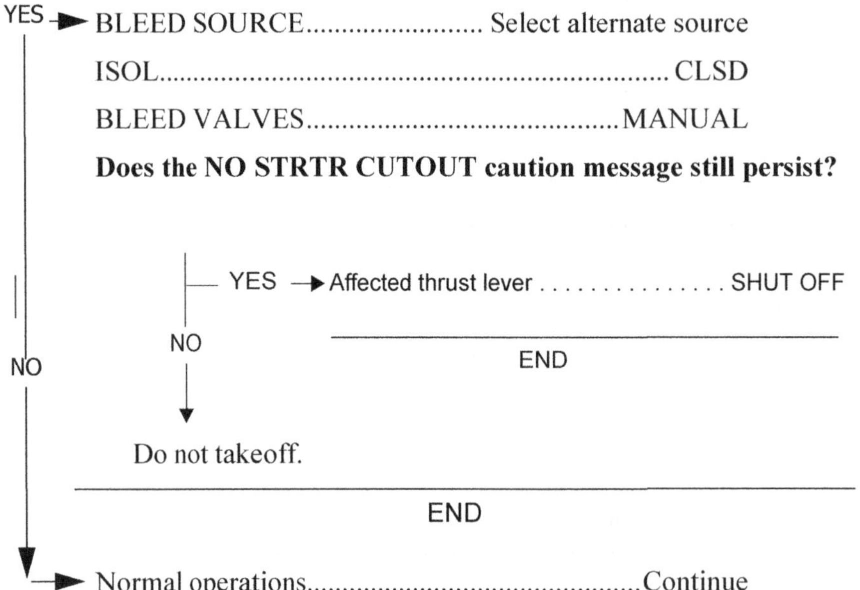

CONTINUED ON NEXT PAGE

EXTERNAL AIR START:

Affected ENG STOP ....STOP

Does the NO STRTR CUTOUT caution message persist?

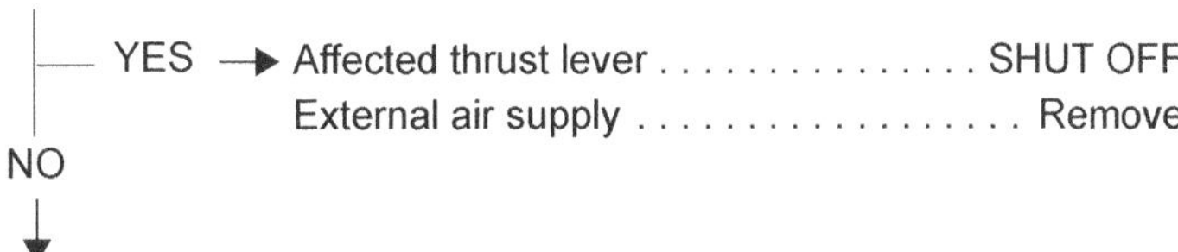

Normal operations ....Continue

NO START BEING ATTEMPTED:

Affected Thrust Lever.....................................SHUT OFF

L (R) STRT VLV OPEN (In flight)

Choose a scenario

APU or CROSS BLEED START:

Affected ENG STOP ....STOP

Does the L (R) STRT VLV OPEN caution message persist?

YES → BLEED SOURCE....... Select non-affected engine source

ISOL... CLSD

BLEED VALVES...MANUAL

Does the L (R) STRT VLV OPEN caution message still persist?

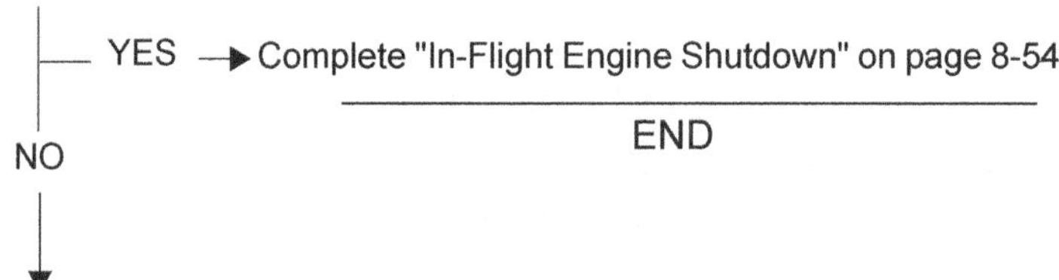

NO

Inoperative PACK...OFF

NOTE:
Airplane altitude not above 25,000 feet during single pack operations with the engine as the bleed source.

WING A/I CROSS BLEED...... Select source engine side

ANTI-ICE, LH or RH COWL (Non-source engine side). ..... OFF

Leave icing conditions to prevent ice accumulation on inoperative cowl.

CONTINUED ON NEXT PAGE

NOTE:
Icing conditions exist in flight at a TAT of 10°C (50°F) or below and visible moisture in any form is encountered (such as clouds, rain, snow, sleet or ice crystals), except when the SAT is -40°C (-40°F) or below.

NO

END

→ Normal operations ....Continue

L (R) STRT VLV OPEN (On the Ground)

Choose a scenario

APU or CROSS BLEED START:

Affected ENG STOP ....STOP

Does the L (R) STRT VLV OPEN caution message persist?

YES → BLEED SOURCE....... Select non-affected engine source

ISOL.. CLSD

BLEED VALVES...MANUAL

Does the L (R) STRT VLV OPEN caution message still persist?

YES → Affected thrust lever . . . . SHUT OFF

END

NO ↓

Do not take off

END

NO → Normal operations...Continue

CONTINUED ON NEXT PAGE

EXTERNAL AIR START:

Affected ENG STOP ....STOP

Does the L (R) STRT VLV OPEN caution message persist?

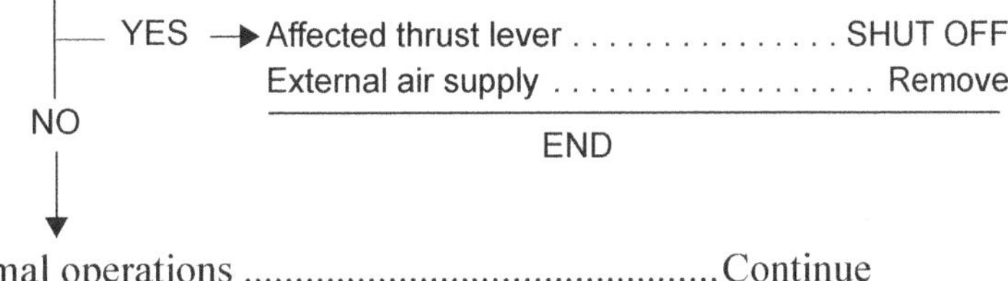

Normal operations ....Continue

NO START BEING ATTEMPTED:

Affected thrust lever ....SHUT OFF

L (R) ENG SRG CLSD

Choose a scenario

IN FLIGHT:

Affected thrust lever .... Adjust as required
Avoid rapid acceleration or decelerations.
Avoid high thrust settings.

Did compressor stall occur?

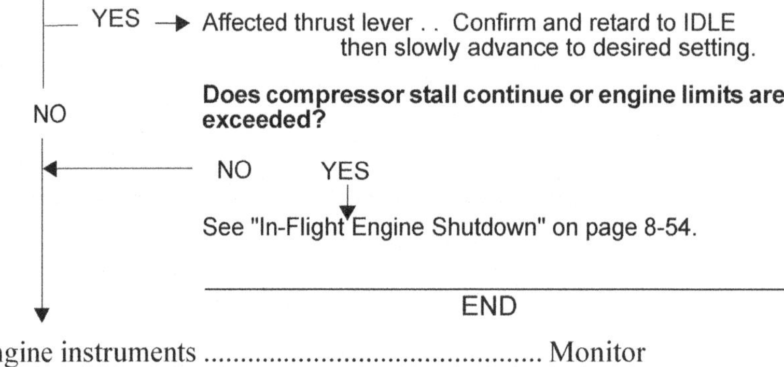

Engine instruments .... Monitor

CAUTION:
If required, use remaining thrust reverser carefully upon landing.

ON GROUND:

Do not takeoff.

L (R) ENG SRG OPEN

Choose a scenario

IN FLIGHT:

Affected engine.. Monitor ITT and adjust thrust to maintain ITT limits

Do not deploy the affected thrust reverser upon landing.

ON GROUND:

Do not takeoff.

ENG BTL 1 (2) LO

No action required.

N_1 Fan Vibration

CAUTION:
It is not recommended that an engine be shut down unless there is another indication of a severe engine abnormality (e.g., high oil temperature, high oil pressure, or abnormal vibration felt throughout the airframe).

Are icing conditions present or is ice accumulation on the fan suspected?

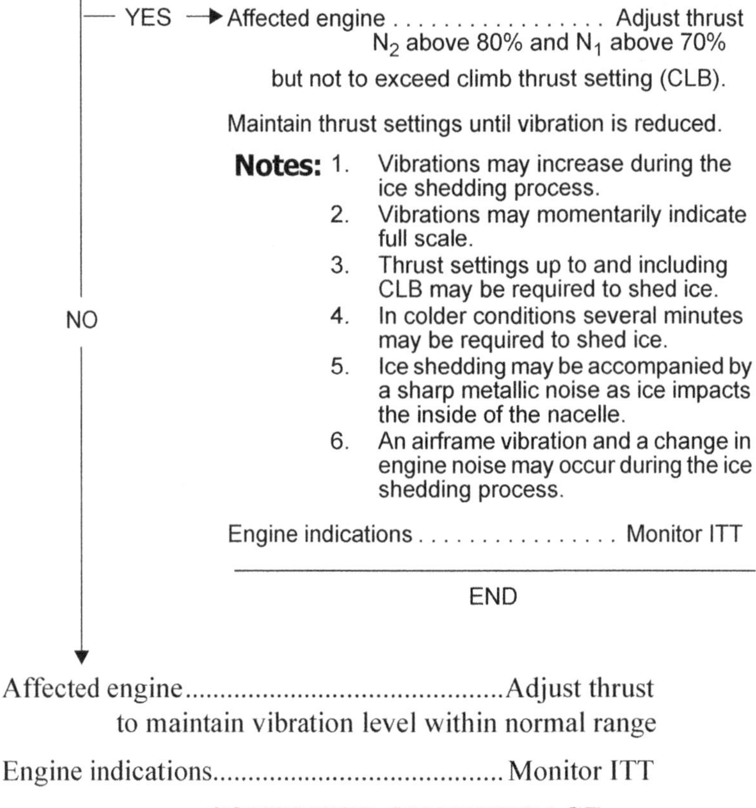

CONTINUED ON NEXT PAGE

Can the vibration be controlled (reduces with thrust reduction) or reduced to within normal operating range?

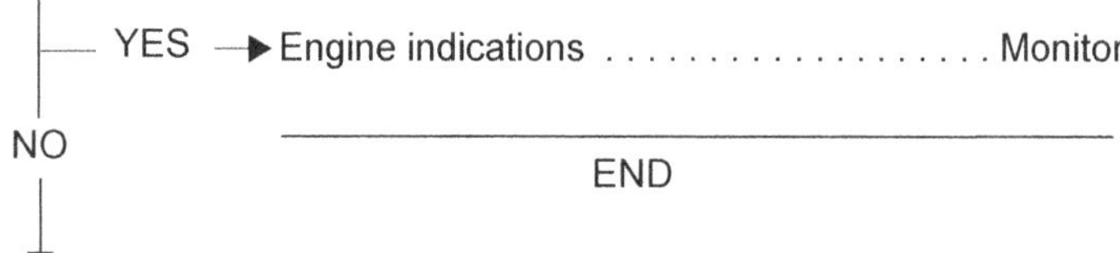

Are one or more additional signs of abnormal engine operation present (high oil temperature or pressure, etc.)?

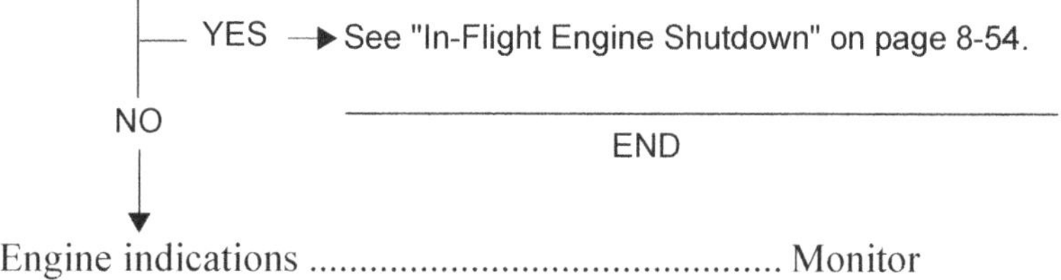

Engine indications .... Monitor

N_2 Core Vibration

CAUTION:
It is not recommended that an engine be shut down unless there is another indication of a severe engine abnormality (e.g., high oil temperature, high oil pressure or abnormal vibration IS felt through the airframe).

Affected engine..Adjust thrust
to maintain vibration levels within normal range

Can the vibration be controlled (reduces with thrust reduction) or reduced to within normal operating range?

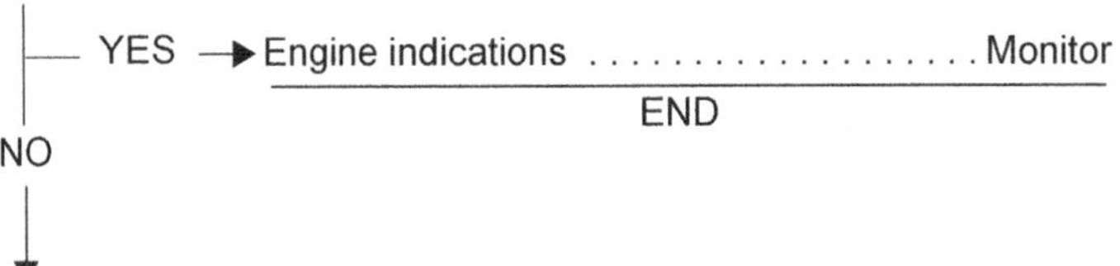

Are one or more additional signs of abnormal engine operation present (high oil temperature or pressure, etc.)?

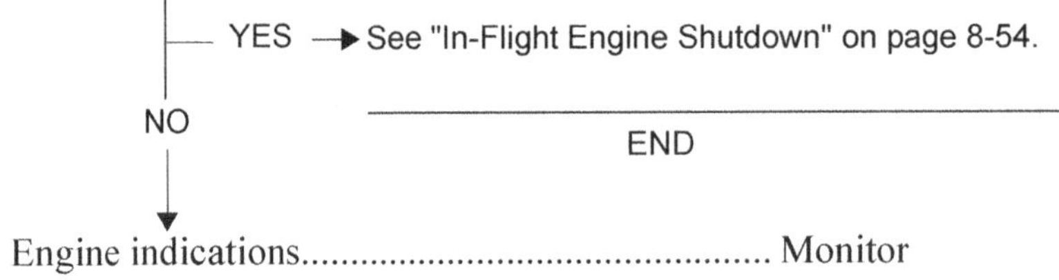

Engine indications... Monitor

Engine Oscillations

ENG SYNC .... OFF

Does the engine stabilize?

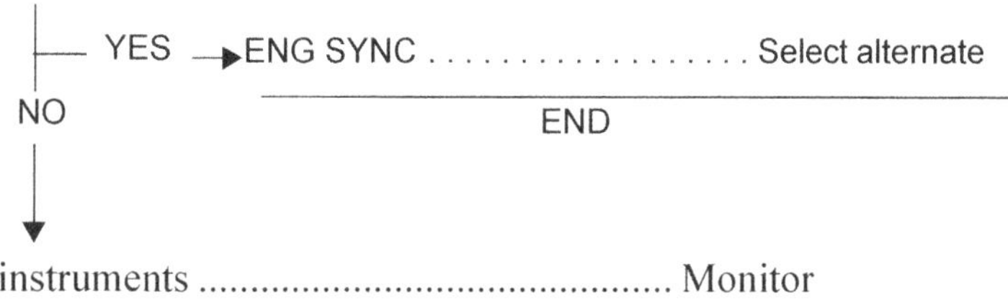

Engine instruments .... Monitor

L (R) ENG DEGRADED

Engine instruments .... Monitor

NOTE:
Engine performance/operation may be degraded.

NOTE:
The L and R ENG DEGRADED caution messages are displayed simultaneously after wing anti-ice selection, then monitor for higher than normal ITTs at CLB thrust setting and above. Thrust reduction will reduce ITTs to normal values. In this condition, flight with wing anti-ice on may continue.

High Oil Temperature

Readout of 156 to 163°C (amber range)

Readout of 164°C and above (red range)

Affected engine..Adjust thrust

to maintain oil temperature within normal range

Can oil temperature be reduced below red range?

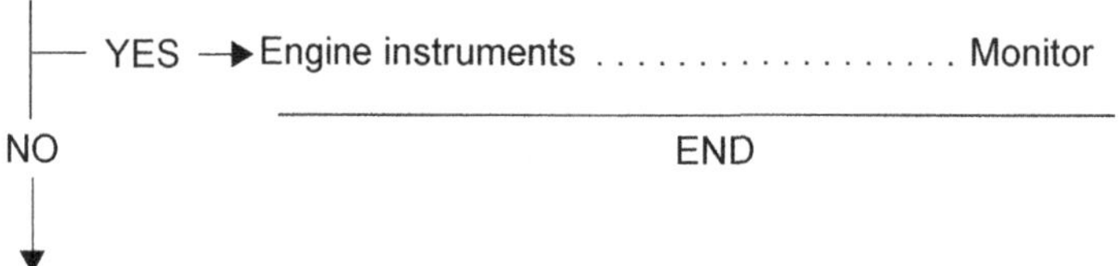

See "In-Flight Engine Shutdown" on page 8-54.

Hot Start

Affected engine, thrust lever...........................SHUT OFF

Dry Motor ....Until ITT less than 120°C
or starter limit, whichever comes first

Affected engine, ENG STOP..................................STOP

L (R) START VALVE

Choose a scenario

IN FLIGHT:

Windmilling Relight Procedure..................... Accomplish

See “Windmilling Relight” on page 8- 60.

ON GROUND:

Affected engine, ENG STOP...................................STOP

Air turbine start is not available.

L (R) ENG TAT HEAT

Avoid icing conditions. Affected engine T2 probe is not heated.

Loss of FAN VIB Indicator

Monitor fan vibration indicator on operational side.

Avoid icing conditions.

APU LCV OPEN

Is the APU GEN required?

YES →

BLEED SOURCE.....................R ENG

ISOL .... CLSD

BLEED VALVES....................MANUAL

Note: Disregard the APU BLEED ON caution message, if present.

L PACK ....OFF

WING A/I CROSS BLEED........FROM RIGHT

LH COWL ANTI-ICE ....OFF

Leave icing conditions to prevent ice accumulation on inoperative engine cowl.

Note: Icing conditions exist in-flight at a TAT of 10°C (50°F) or below, and visible moisture in any form is encountered (such as clouds, rain, snow, sleet or ice crystals), except when the SAT is -40°C (-40°F) or below.

END

NO ↓

BLEED SOURCE.. BOTH ENG

ISOL.. CLSD

BLEED VALVES..MANUAL

NOTE:

Disregard the BLEED MISCONFIG caution message.

APU START/STOP..Select off

NOTE:

The APU is available for restart, if the RPM is stabilized at 0.

APU RPM.. Monitor

CONTINUED ON NEXT PAGE

Is the APU RPM more than 0?

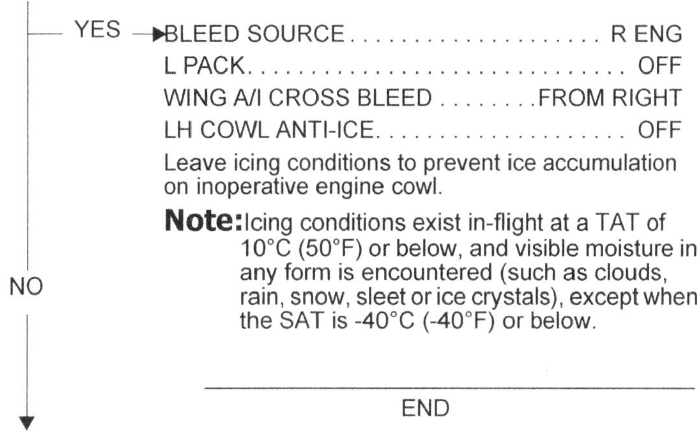

No further action required.

APU LCV CLSD

BLEED SOURCE.................................... Select alternate source if required

Manual Bleed Procedure

(to select both engines)

As the bleed source .... Accomplish

See “Manual Bleed Procedure” on page 8- 247.

APU BLEED ON

Does APU BLEED ON caution message occur during normal ECS operation?

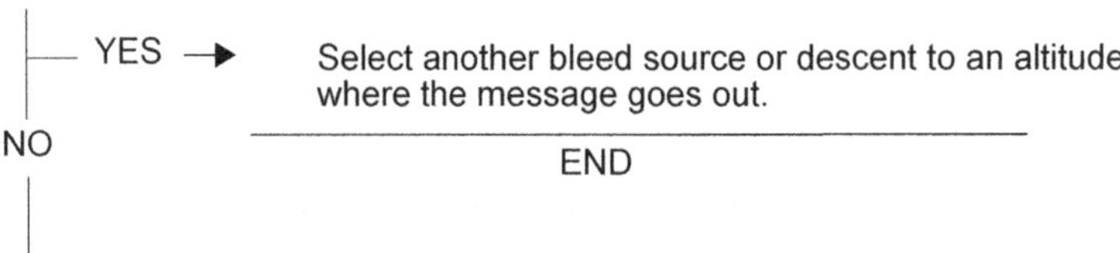

If the APU BLEED ON caution message occurs during a "Starter-Assisted APU Bleed Relight", no further action is required.

APU FAULT

Is APU generator required?

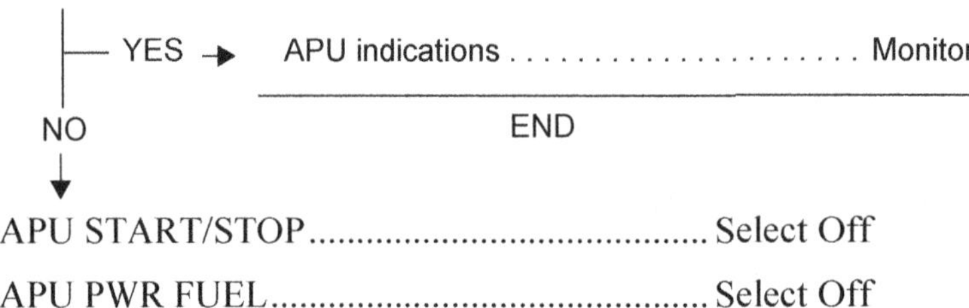

APU START/STOP..Select Off

APU PWR FUEL...Select Off

APU PUMP

Is APU generator required?

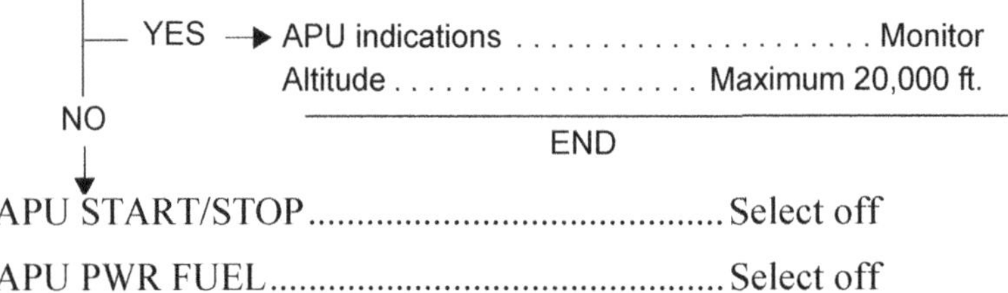

APU START/STOP..Select off

APU PWR FUEL...Select off

APU ECU FAIL

Is APU generator required?

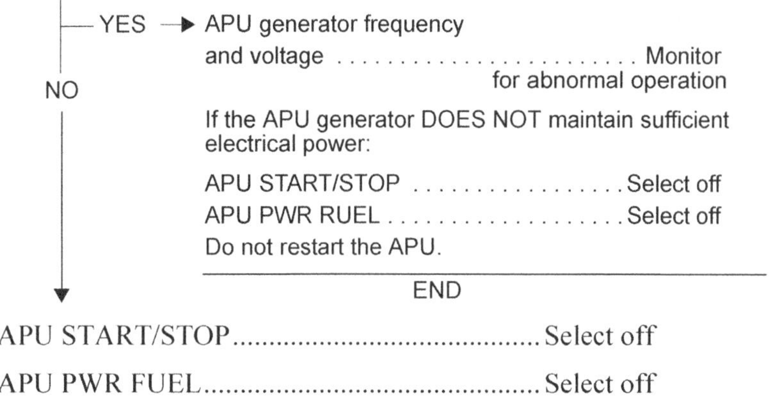

APU START/STOP.. Select off

APU PWR FUEL.. Select off

APU SOV FAIL

APU START/STOP.. Select off

APU PWR FUEL.. Select off

APU SOV OPEN

APU START/STOP.. Select off

APU PWR FUEL.. Select off

APU Door Failure

Choose a scenario

APU DOOR OPEN caution message (primary display) and DOOR OPEN amber message (STAT page):

Airspeed .... Not more than 220 KIAS

NOTE:

The APU may be restarted and operated throughout the remainder of the flight without any airspeed restriction.

DOOR INHIB/CLSD:

APU START/STOP .... Select off

APU PWR FUEL .... Select off

DOOR INHIB/OPEN:

Is the APU operating?

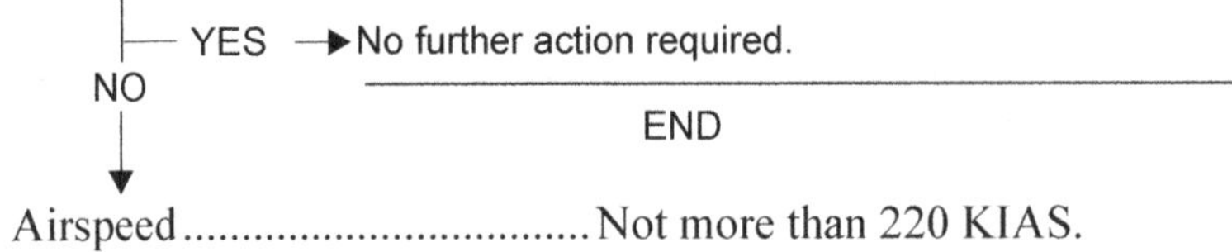

Airspeed .... Not more than 220 KIAS.

CONTINUED ON NEXT PAGE

DOOR INHIB - - - - - or DOOR - - - - -:

Is the APU operating?

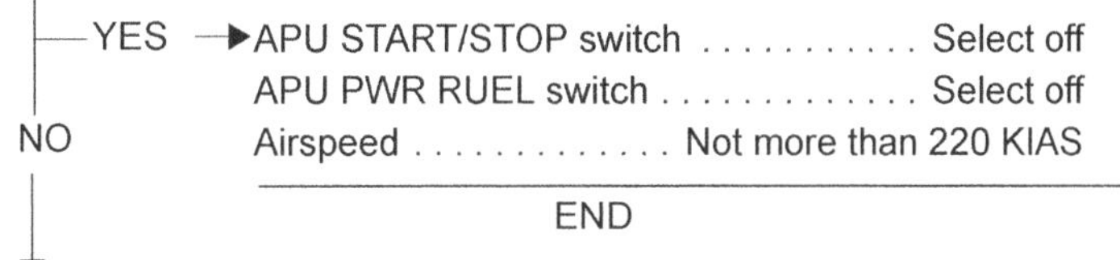

Do not attempt to start the APU.

Airspeed..................................Not more than 220 KIAS.

APU FIRE

APU FIRE PUSH switch....................Confirm and select

After 5 seconds and APU FIRE warning message persists:

APU BOTTLE switch...Select

CAUTION:

Do not restart the APU.

APU START/STOP switch ....Select off

APU PWR FUEL switch ....Select off

Land immediately at the nearest suitable airport.

APU OVERSPEED

APU START/STOP switch .... Select off

APU PWR FUEL switch .... Select off

CAUTION:
Do not restart the APU.

APU OVERTEMP

Choose a scenario

On the ground:

APU START/STOP switch .... Select off

APU PWR FUEL switch .... Select off

CAUTION:
Do not restart the APU.

In flight:

Is APU GEN required?

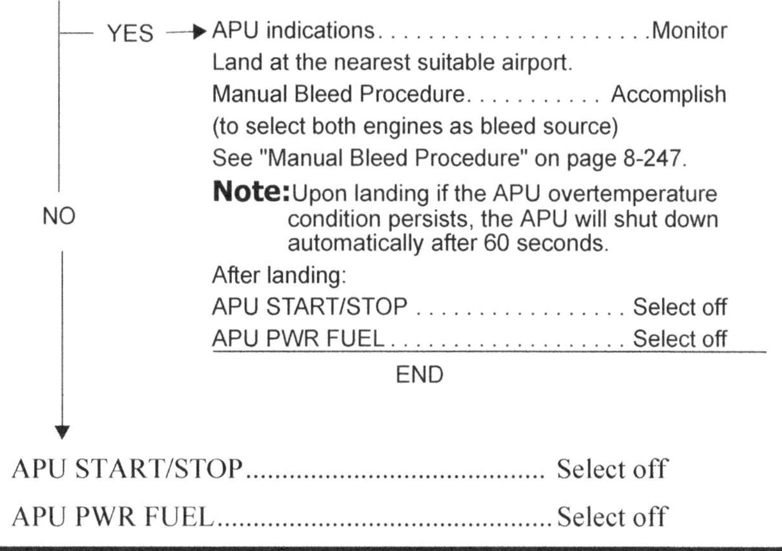

APU START/STOP... Select off

APU PWR FUEL... Select off

APU FIRE FAIL

Is the APU generator required?

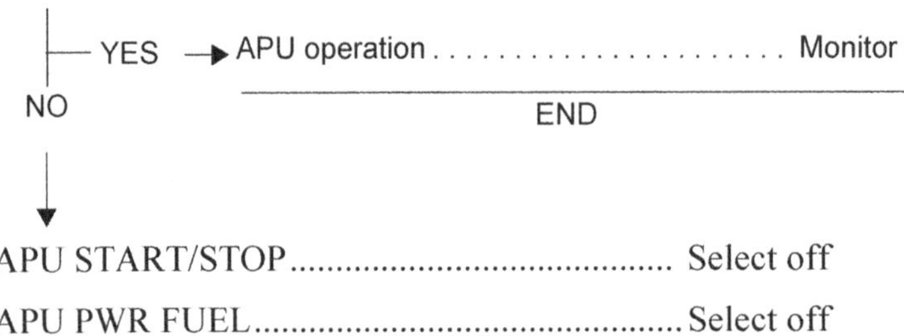

APU START/STOP.. Select off

APU PWR FUEL... Select off

APU BTL LO

Is a fire detected?

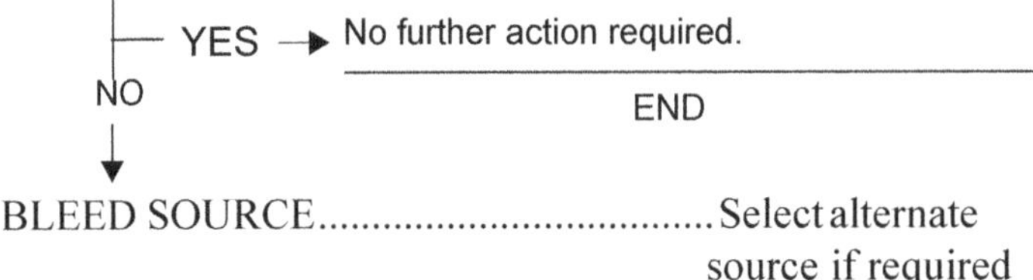

BLEED SOURCE..................................... Select alternate source if required

Is the APU generator required?

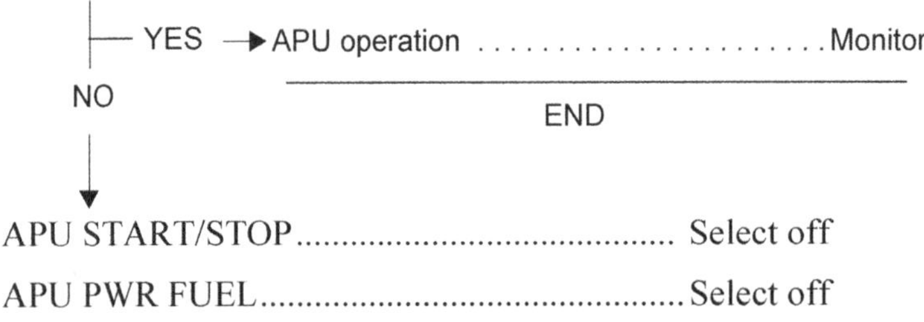

APU START/STOP.. Select off

APU PWR FUEL... Select off

Engine Failure During Approach

CEME: N11

Autopilot (if engaged).................................... Disengage

Operating Engine .... Increase thrust as required

Flight Spoilers (if extended) .... Retract

Approach and landing flaps.................................. Use 20°

Airspeed.......... Increase to V_{REF} (FLAPS 45) + 12 KIAS

Airplane .... Retrim and continue approach, or go-around at pilot's discretion

NOTE:
The autopilot may be re-engaged, if above 400 feet AGL.

If continuing the approach:

GRND PROX, FLAP witch................................... OVRD

Final approach speed .... Maintain V_{REF} (FLAPS 45) + 12 KIAS

Actual landing distance....................................... Increase

Without two Thrust Reversers and/or Wet/Contaminated Runway Surface	With two Thrust Reversers and a Dry Runway Surface
1.30 (30%)	1.25 (25%)

CAUTION:
If required, use remaining thrust reverser carefully upon landing.

After landing:

Affected thrust lever ....SHUT OFF

Intentionally Left Blank

Electrical Index

Intentionally Left Blank

Electrical

EMER PWR ONLY (ADG is deployed)

STAB TRIM CH 2............................... Confirm engaged

Engine instruments .... Verify N_1, N_2, and ITT

Have both engines failed?

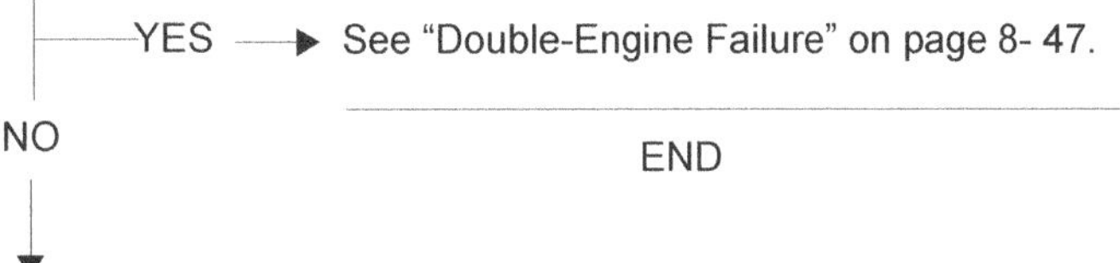

Cabin altitude... Monitor

GEN 1 and GEN 2Confirm and OFF/RESET then AUTO

APU (if available 37,000 feet and below) ....START

Did any generator come on-line?

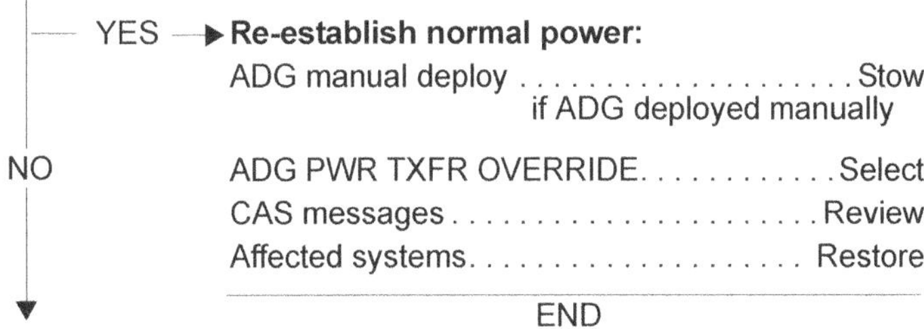

Continue flight on emergency power only:

R PACK ....Confirm on

L PACK ....OFF

PRESS CONTROL.. MAN

Leave icing conditions.

NOTE:
Icing conditions exist in-flight at a TAT of 10°C (50°) of below, and visible moisture in any form is encountered (such as clouds, rain, snow, sleet or ice crystals), except when the SAT is -40°C (-40°F) or below.

CONTINUED ON NEXT PAGE

Land at the nearest suitable airport.

Avoid excessive rudder inputs.

The following significant systems are not available when on emergency power only:

- Automatic pressurization
- WIndshield wipers, both windshield heaters, right window heater
- Yaw damper 2 and autopilot
- All ground spoilers and inboard multi-function spoilers
- Stabilizer trim channel 1, aileron and rudder trims
- Hydraulic pumps 1B, 2B, and 3A
- Copilot's PFD, MFD, navigation and communications systems
- Copilot's instrument lights, NAV lights, and taxi lights
- Right probe heaters and ice detector
- Anti-skid system and nosewheel steering
- Below 135 KIAS, AC ESS BUS is shed causing the loss of the following:
  - The remaining TRU (ESS TRU 1 or ESS TRU 2, as applicable)
  - Rudder limiter
  - XFLOW pump
  - Left probe heaters and ice detector 1
  - Left window heat

During Approach - Prior to reducing speed below 145 knots:

LDG GEAR lever ....DN

FLAPS ....Set for landing

CONTINUED ON NEXT PAGE

Actual landing distance is increased by a factor of:

Without two Thrust Reversers and/or Wet/ Contaminated Runway Surface	With two Thrust Reversers and a Dry Runway Surface
2.30 (130%)	1.80 (80%)

NOTE:
The slats/flaps will operate at half speed:

NOTE:
A momentary loss of ADG power may occur:

- At 140 KIAS and below, if the slats/flaps are operating
- At 108 KIAS and below, if pitch trim is used.

Upon landing:

NOTE:
Use the thrust reversers as required during landing.

Do not cycle the brakes.

Brake pedals.. Depress
apply slowly and steadily

Loss of All AC Power (ADG not operative)

ADG manual deploy handle .... Pull

Did the ADG deploy and now supplying power?

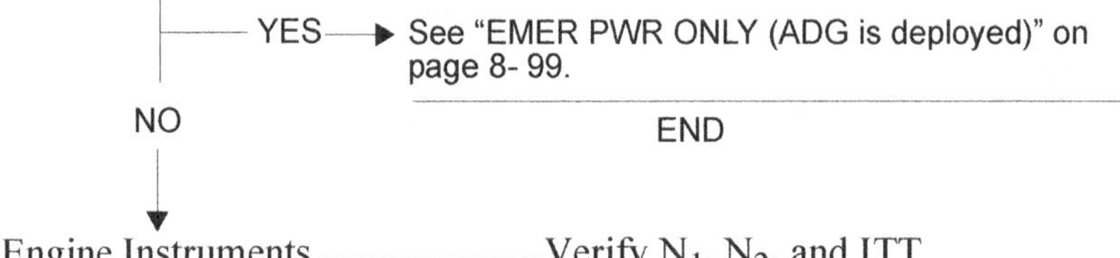

Engine Instruments .... Verify N_1, N_2, and ITT

Do engine instruments indicate the engines are operating?

YES → Cabin altitude .... Monitor
GEN 1 and GEN 2 .... Confirm OFF/RESET then AUTO
APU (if available, at 37,000 feet and below) .... START
NOTE: Each APU start attempt consumes approximately six minutes of battery life.

NO ↓

Double Engine Failure procedure. . . . . Accomplish
See "Double-Engine Failure" on page 8-47.

Did any generator come on line?

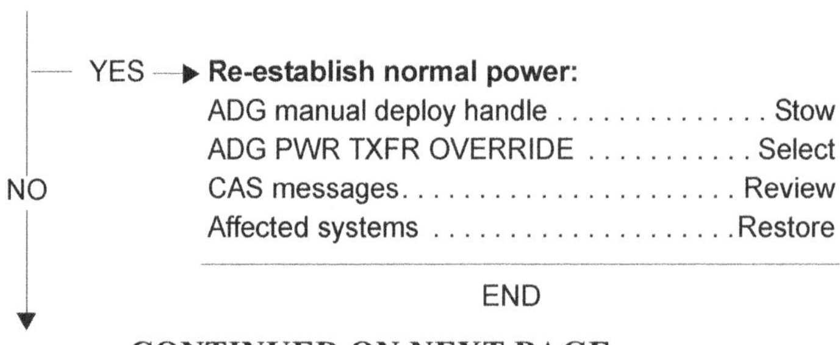

CONTINUED ON NEXT PAGE

Continue flight on battery power only:

NOTE:
Electrical power may be lost after 30 minutes.

STALL PTCT PUSHER

(left or right) .... OFF

R PACK .... Confirm on

L PACK .... OFF

PRESS CONTROL .... MAN

Leave icing conditions.

NOTE:
Icing conditions exist in-flight at a TAT of 10°C (50°) of below, and visible moisture in any form is encountered (such as clouds, rain, snow, sleet or ice crystals), except when the SAT is -40°C (-40°F) or below.

Land at the nearest suitable airport.

The following significant systems are not available when on battery power only:

- Rudder limiter
- Automatic Pressurization
- WIndshield wipers, both windshield heaters, and window heaters
- Yaw damper 2 and autopilot
- All ground spoilers and all multi-function spoilers
- Stabilizer, aileron and rudder trims
- Hydraulic pumps 1B, 2B, 3A and 3B
- Copilot's PFD, MFD, navigation and communications systems
- Copilot's instrument lights, NAV lights, and taxi lights
- Probe heaters and ice detectors

CONTINUED ON NEXT PAGE

- Anti-skid system and nosewheel steering
- Normal landing gear extension
- Flaps and slats
- All TRUs
- XFLOW pump

Prior to landing:

Landing Gear Manual Extension .... Accomplish

See "Landing Gear Manual Extension" on page 8- 298.

Final Approach speed .... $V_{REF\ (FLAPS\ 45°)} + \Delta V_{REF}$ from the following table

CONTINUED ON NEXT PAGE

ΔV_{REF}			
Flaps Position	**Slats Position**		
	0-19	**20-24**	**25**
0-7	40	24	24
8-19	30	18	18
20-29	30	12	12
30-44	24	24	8
45	10	10	0

Actual landing distance is increased by a factor of:

Final Approach Speed ΔV_{REF} (KTS)	**Actual Landing Distance Factor (Without two Thrust Reversers and/or Wet/Contaminated Runway Surface)**	**Actual Landing Distance Factor (With two Thrust Reversers and a Dry Runway Surface)**
40	4.00	2.75
30	3.50	2.50
24	3.10	2.40
18	2.95	2.25
12	2.80	2.05
10	2.70	2.05
8	2.65	2.00
0	2.35	1.80

CAUTION:

The maximum tire speed (195 knots ground speed) may be exceeded at high oat and/or high airfield elevation.

CONTINUED ON NEXT PAGE

After touchdown:

Do not cycle the brakes.

Brake pedals.. Depress
apply slowly and steadily

NOTE:
A slight pitch-up tendency may occur upon selection of reverse thrust. This can be readily corrected by the application of nose-down elevator and/or brakes.

AC 1 (2) AUTO XFER

Is the affected AC bus powered?

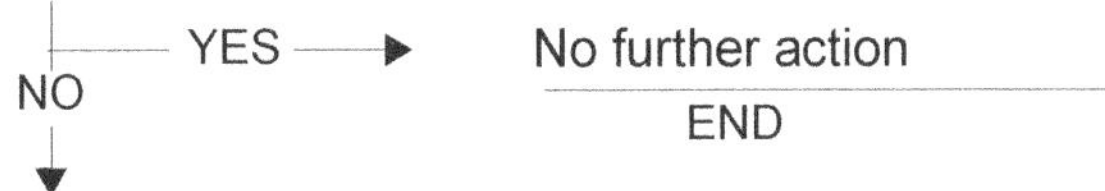

Affected airplane systems.................................... Review

AC Bus 1	AC Bus 2
Hydraulic Pump 2B and 3B	Hydraulic Pump 1B and 3A
Avionics Display Cooling Fan 2	Pitch Trim 2
Cabin Feeder 1	Slats 2
Pitch Trim 1	Flaps 2
Slats 1	Recirc Fan 2
Flaps 1	Right Windshield Heater
Recirc Fan 1	Ice Detector 2
Hydraulic System Fan	TRU 2
Left Windshield Heater	Galley Heater
TAT Probe Heater	Integral Lighting Copilot Panels
Right AOA Heater	Right Window Heater
Right Pitot Heater	ESS TRU 2
Lav Exhaust Fan	Galley Exhaust Fan
Ground Prox Warning	QAR
Main Battery Charger	IRU Fan
Engine Vibration Monitor	Printer
Flight Recorder Power	
TRU 1	
Baggage Compartment Heater	
Air Driven Generator Heater	

AC BUS 1

GEN 1 .... Confirm and OFF/RESET then AUTO

Does the AC BUS 1 caution message disappear?

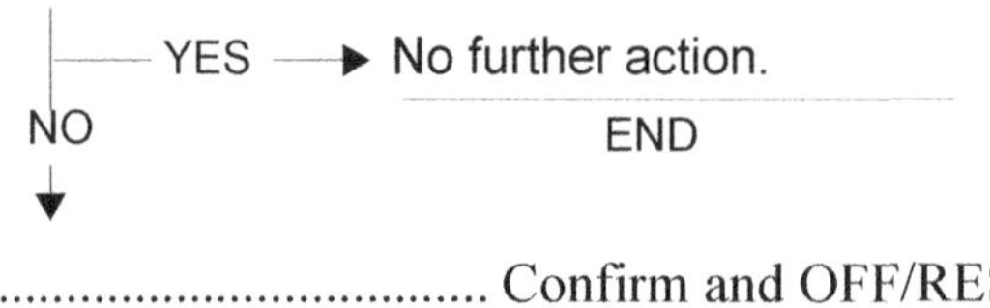

GEN 1 .... Confirm and OFF/RESET

APU (if available, 37,000 feet and below) ....START

If the AC BUS 1 caution message still persists:

Affected airplane systems..................................... Review

AC Bus 1	
Hydraulic Pump 2B and 3B	Avionics Display Cooling Fan 2
Cabin Feeder 1	Lav Exhaust Fan
Pitch Trim 1	Ground Prox Warning
Slats 1	Main Battery Charger
Flaps 1	Engine Vibration Monitor
Recirc Fan 1	Flight Recorder Power
Hydraulic System Fan	TRU 1
Left Windshield Heater	Baggage Compartment Heater
TAT Probe Heater	Air Driven Generator
Right AOA Heater	
Right Pitot Heater	

No further action required.

AC BUS 2

GEN 2 .... Confirm and OFF/RESET then AUTO

Does the AC BUS 2 caution message disappear?

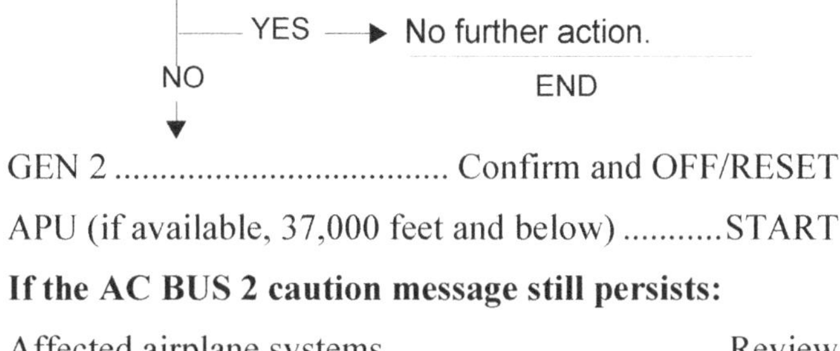

GEN 2 .... Confirm and OFF/RESET

APU (if available, 37,000 feet and below) ....START

If the AC BUS 2 caution message still persists:

Affected airplane systems..................................... Review

AC Bus 2	
Hydraulic Pump 1B and 3A	Galley Heater
Pitch Trim 2	Integral Lighting Copilot Panels
Slats 2	Right Window Heater
Flaps 2	QAR
Recirc Fan 2	ESS TRU 2
Right Windshield Heater	Galley Exhaust Fan
Ice Detector 2	IRU Fan
TRU 2	Printer

No further action required.

AC SERV BUS

Affected airplane systems .... Review

AC SERV BUS	
AC SERVICE Feed	APU Charger
Toilet	Cabin Lighting Sidewall
Water System	Logo Lights
Cabin Lighting Ceiling	

AC ESS BUS

AC ESS XFER... ALTN

Does the AC ESS BUS caution message disappear?

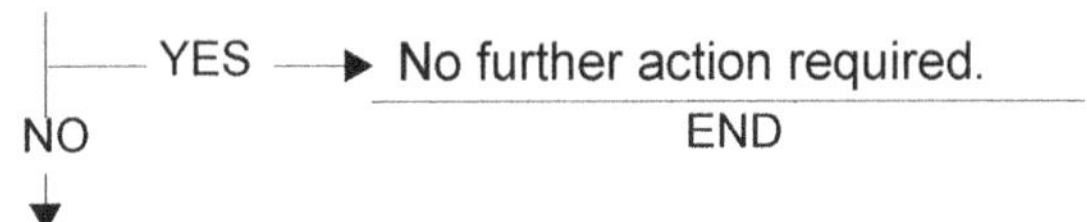

Affected airplane systems .... Review

AC ESS BUS	
Crossflow Pump	Engine Ignition A
Essential TRU 1	Display Cooling Fan 1
Left Pitot Heater	Integral Lighting CB Panels
Left AOA Heater	Integral Lighting Pilot Panels
Standby Pitot Heater	Integral Lighting Center Panels
Cabin Lighting Ceiling	Integral Lighting Overhead Panels
Ice Detector 1	TCAS
Avionics Display Cooling Fan 1	Left Window Heater

No further action required.

APU BATT OFF

BATTERY MASTER ....ON

Does the APU BATT OFF caution message disappear?

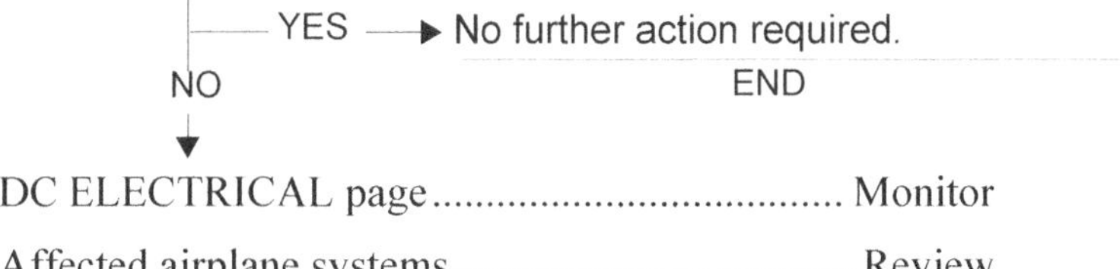

DC ELECTRICAL page .... Monitor

Affected airplane systems .... Review

AC BATT DIRECT BUS	
APU Battery Power Sensors	APU Door Actuator
APU Battery Control	Engine Oil Indicator
DCPC back-up power 1	Refuel/Defuel Panel
APU ECU Secondary	Emergency Refuel
External AC back-up power	EMER BUS Feed
SERV BUS Feed	Engine Oil Replenishment

NOTE:
APU start is not possible.

NOTE:
The APU shutdown sequence may be altered. Wait 30 seconds before selecting APU PWR/FUEL to OFF or the BATTERY MASTER to OFF (the 30 second wait allows the APU ECU to log all events). The APU door may not close.

APU GEN OFF

APU GEN .... Confirm and OFF/RESET then AUTO

AC ELECTRICAL page .... Monitor

Does the APU GEN OFF caution message still persist?

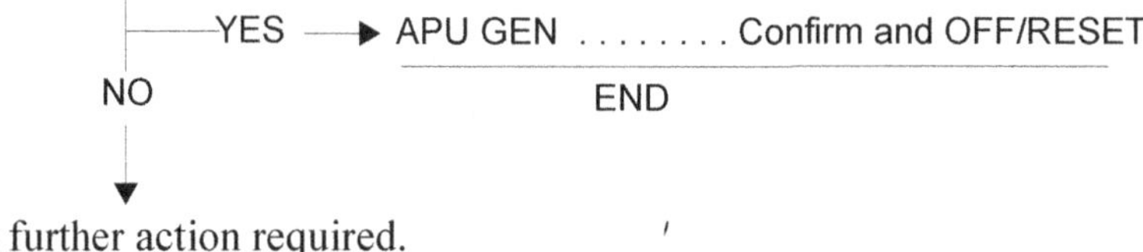

No further action required.

APU GEN OVLD

AC ELECTRICAL page .... Monitor

AC loads .... Reduce as necessary

BATTERY BUS

Affected airplane systems.................................... Review

Airplane altitude .... Not more than 13,000 feet

BATTERY BUS	
ACS Control 1 Channel A	Stick Pusher
Flaps Control Channel 2	Stall Protection Left Channel
Slats Control Channel 2	FADEC L & R back-up power
Aileron/Rudder Trim Indication	Engine Ignition B
Hydraulic System Indications 3	VHF COM 1
Emergency Bus Feed	Emergency Tuning
Left Engine Oil Pressure Indication	Pilot / Copilot / Observer Audio
Left & Right Engine Start	Generator Control Units 1, 2, & 3
Left Fuel Pump	ACPC Control 3
Left Fuel Pump Control	BATT BUS Feed
Cargo Smoke Detector A & B	L & R Engine Anti-Ice Valves
Passenger signs	Wing Anti-Ice Isolation Valve
Fuel System Control	ADG Deploy auto & Manual
Fire Detection A and B	MLG Gay Overheat Detection
APU Control	Integrated Standby Instrument
Fuel Gravity Crossflow	Clock 1
Left Transfer Fuel SOV	PSEU Channel A & B
APU Fuel Pump	WOW Relay
APU ECU Primary	RAM Air SOV
Wing Landing Lights	Cabin Pressure Manual Control
Pilot / Copilot / Observer Map Lights	EMER DEPRESS
Cabin Utility Lights	Left IAPS AFCS / MDC
Overhead Panel Lights	Left IAPS Fan
EICAS / RTU Dimming	Right Transfer Fuel SOV
Pass OXY Manual Deploy L & R	Passenger Oxygen Auto Deploy L & R
IDG 1 and 2 Disconnect	L & R Crew Oxygen Monitor
EICAS DCU 1 & 2	EICAS Bright / Dim Power 1 & 2
Cabin Interphone	EICAS Primary Display

CONTINUED ON NEXT PAGE

BATTERY BUS	
Passenger Address	EICAS Secondary Display
Right Lamp Driver Unit	EICAS Control Panel

Land at the nearest suitable airport.

Prior to landing:

L and R PACKs .... OFF

DC BUS 1

Affected airplane systems.................................... Review

DC BUS 1	
Anti-Ice Control Channel A	Wing Inspection Lights
Baggage Comp Cont	Maintenance Lights
Lav Smoke Detection	GPS 1
ACPC Control 1	Left Windshield Heater Control
Passenger Door Actuator	ADS Heater Control 2
L Fwd/Aft Cabin Reading Lights	Right Static Heater
Flight Recorder Control	Left IAPS (or IAPS/FMS)
SSCU 1 Channel A	EICAS Primary Display
P FEEL 1 RTL 1	EICAS Secondary Display
Left T2 Heater	Left Lamp Driver Unit
Taxi Lights	EICAS Bright/Dim Power 1 and 2
Fan Monitor	CDU 1
Cabin Pressure Control 1	Data Loader
Hydraulic AC Pump Control 3B	DME 1
Hydraulic Fan Control	AFT Cabin Temperature Sensor
Hydraulic Indicator 2	Radar Altimeter 1
Hydraulic AC Pump Control 2	WX Radar R/T and Control 1/2
PSEU Channel A	ACS Control 2 Channel A
Rear Anti-Collision Lights	
Nose Steering	Brake Pressure Application
Cockpit Dome Light	Cockpit Temperature Sensor
Boarding Music	Wiper, Pilot side
Anti-Skid	Cockpit Floor Lights
Nose Landing Lights	

DC BUS 2

Affected airplane systems.................................... Review

DC BUS 2	
Fwd Cabin Temperature Sensor	Anti-Skid
Radar Altimeter 2	Wiper, Copilot
EICAS DCU 2	Chart Holder Lights
ACS Control 1 Channel B	Copilot Map Lights
PFD 2	Wing Anti-Collision Lights
MFD 2	Right Fuel Pump
EFIS Control Panel 2	Right Fuel Pump Control
Right ACS Manual	GPS 2
RTU 2	Right Windshield Heater Control
SSCU 1 Channel B	Right Window Heater Control
Rudder Trim	Right IAPS
Aileron Trim	Right IAPS AFCS
Right T2 Heater	Right IAPS Fan
Service Bus Feed	Clock 2
Left ACS Pressure Sensor	ADC 2
Cabin Pressure Control 2	ADF 2
Galley Heater Control	Transponder 2
Hydraulic Indicator 1	CDU 2
Hyd. AC Pump Control 1/3A	VHF NAV and COM 2
PSEU Channel B	Nose Steering
ACPC Control 2	DME 2
Brake Pressure Indicator	Observer's audio

DC EMER BUS

Affected airplane systems .... Review

DC EMER BUS	
FIREX A	Right Engine Fuel SOV
FIREX B	Left Engine Fuel SOV
Right Engine Hydraulic SOV	APU Fuel SOV
Left Engine Hydraulic SOV	APU BATT DIR Feed

NOTE:
Engine and APU fire extinguishing systems are inoperative.

NOTE:
APU fuel feed SOVs and ENG fuel SOV will not close.

NOTE:
Engine Hydraulic SOVs will not close.

Land at the nearest suitable airport.

DC SERV BUS

DC Service..ON

Does the DC SERV BUS caution message disappear?

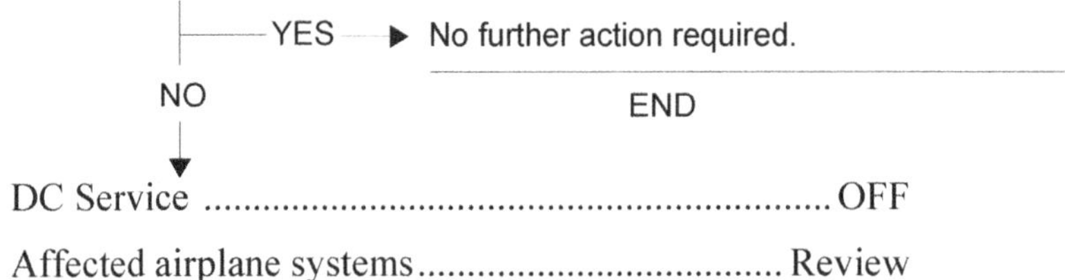

DC Service ....OFF

Affected airplane systems.................................... Review

DC SERVICE BUS	
Forward Service Lights	Galley Area Lights
Aft Service Lights	Service Area Lights
Boarding Lights	Beacon Lights
Navigation Lights	Waste System
Toilet Lights	Water Control

No further action required.

DC ESS BUS

Affected airplane systems.................................... Review

DC ESSENTIAL BUS	
Flaps Control Channel 1	DC ESS Feed
Slats Control Channel 1	Instrument Flood Lights
SSCU 2 Channel A	Emergency Lights Battery Charger
SSCU 2 Channel B	EFIS CRT DImming
PFEEL 2 RTL 2	Stall Protection Right Channel
Crossflow Pump Control	VHF COM 3
Door Indicators	EFIS Control Panel 1
Left Static Heater	EICAS DCU 1
ADS Heater Control 1	RTU 1
ADS Heater Control Standby	Fuel System Control
Left Window Heater Control	Audio Pilot
Thrust Reverser 1/2	ADC 1
Right Eng Oil Pressure Indication	ADF 1
Left Bleed SOV	Transponder 1
Right Bleed SOV	VHF NAV 1
Anti-Ice Control Channel B	Cockpit Voice Recorder
ACS Control 2 Channel B	PFD 1
Left ACS Manual	MFD 1
Display Fan Control	Right ACS Pressure Sensor

Land at the nearest suitable airport

GEN 1 OFF

GEN 1 .... Confirm and OFF RESET, then AUTO

Does the GEN 1 OFF caution message persist?

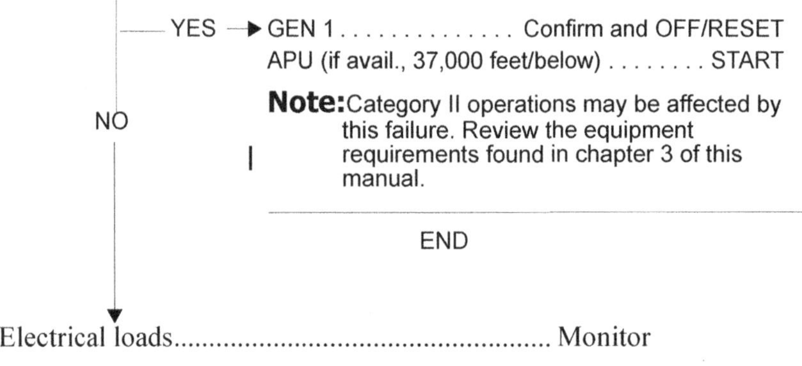

Electrical loads.. Monitor

GEN 2 OFF

GEN 2 .... Confirm and OFF RESET, then AUTO

Does the GEN 2 OFF caution message persist?

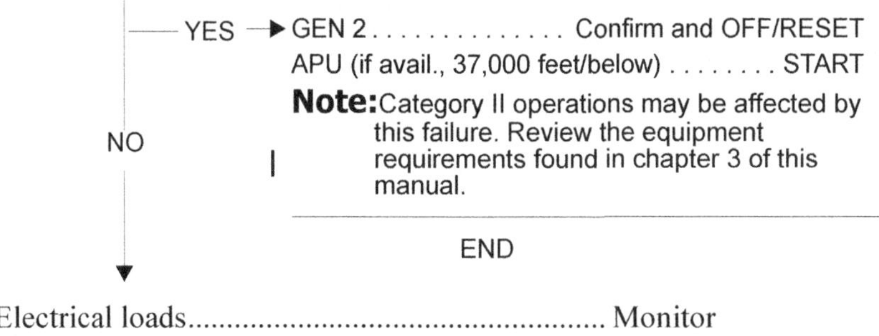

Electrical loads.. Monitor

GEN 1 (2) OVLD

AC ELECTRICAL page............ Monitor GEN 1 (2) load

AC and DC Loads............................ Reduce as necessary

---

IDG 1

GEN 1 .... Confirm and OFF/RESET

IDG 1 DISC ....Confirm and DISC
(for not more than 3 seconds)

NOTE:
The IDG cannot be re-connected during flight.

APU (if available, 37,000 feet and below) ....START

IDG 2

GEN 2 .... Confirm and OFF/RESET

IDG 2 DISC ....Confirm and DISC
(for not more than 3 seconds)

NOTE:
The IDG cannot be re-connected during flight.

APU (if available, 37,000 feet and below) ....START

MAIN BATT OFF

BATTERY MASTER... ON

Does the MAIN BATT OFF caution message disappear?

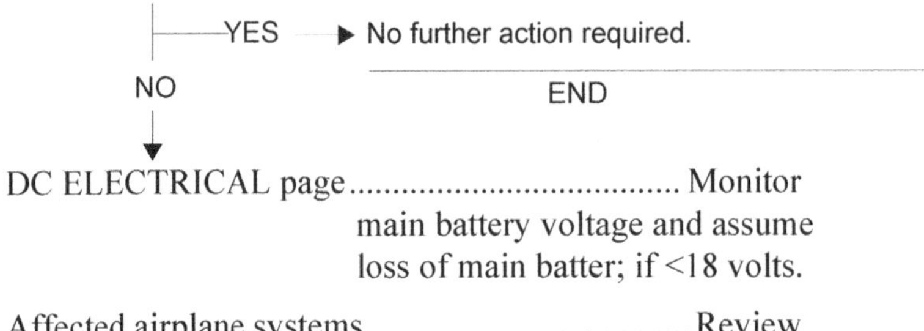

DC ELECTRICAL page...................................... Monitor main battery voltage and assume loss of main batter; if <18 volts.

Affected airplane systems..................................... Review

DC BATTERY DIRECT BUS	
Main Battery Power Sensor	DCPC 2
Main Battery Control	Clock 1
Cockpit Dome Lights	Clock 2
External AC Power 1	Main Battery Charger Output

NOTE:
If main battery voltage is less than 20 volts, both clocks have to be reset.

Smoke, Fire, or Fumes Index

Intentionally Left Blank

Smoke, Fire, or Fumes

WARNING

Time is critical during smoke/fumes/fire emergencies. the flight crew should consider an immediate landing any time the situation cannot be controlled.

WARNING

The flight crew must don oxygen masks / smoke goggles at the first indication of a concentration of smoke or fumes or suspected contaminated air.

CAUTION:
Passenger masks should not be deployed when performing smoke or fire procedures.

CAUTION:
When using the man rate switch to depressurize the cabin, the cabin altitude should not be allowed to exceed 13,500 feet.

CAUTION:
To continue to destination, the flight crew must acknowledge that the threat has been positively identified, confirmed to be extinguished and the smoke/fumes have dissipated.

NOTE:
As part of Standard Operating Procedures for in-flight emergencies, the PASS SIGNS switches are assumed to have been selected ON.

Smoke or Fumes Removal Procedure

WARNING

Accomplish "smoke or fumes removal procedure" if at any time smoke/fumes become the greatest threat or after the source has been eliminated.

Oxygen masks / goggles (if required)............... On, 100%

Crew and cabin communication ....Establish

To stop recirculating smoke/fumes:

AFT CARGO..OFF

RECIRC FAN..OFF

Prepare to land immediately at the nearest suitable airport.

Descent.. Initiate to 10,000 feet or lowest safe altitude

PRESS CONTROL..MAN

MAN ALT..UP

MAN RATE.................................... Increase as required

NOTE:
Make sure that the cabin altitude does not exceed 13,500 feet.

MAN ALT .... HOLD, when smoke has cleared

CONTINUED ON NEXT PAGE

Has the source of smoke/fumes/fire been identified or is the fire controllable?

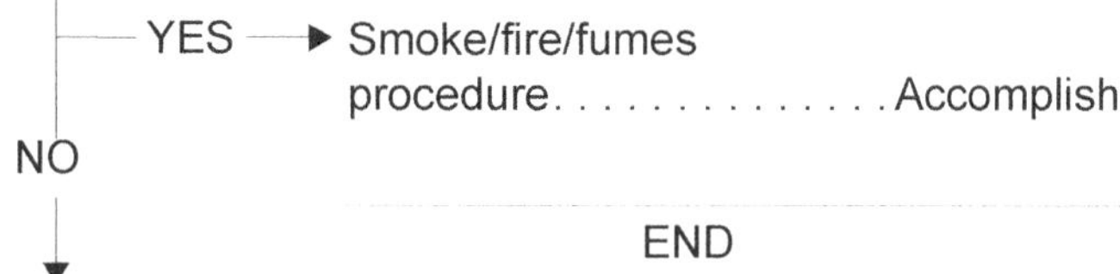

Land immediately at the nearest suitable airport.

CAUTION:
Dependent upon the severity of the situation, the flight crew should expedite the landing. the crew should also consider an overweight landing, tailwind landing, ditching or a forced off-airport landing.

Smoke/Fire/Fumes Procedure

WARNING

Accomplish "smoke or fumes removal procedure" if at any time smoke/fumes become the greatest threat or after the source has been eliminated.

A diversion may be required.

Oxygen masks / goggles (if required).............. On, 100%

Crew and cabin communication ....Establish

To stop recirculating smoke/fumes:

AFT CARGO.. OFF

RECIRC FAN .... OFF

If at any time the smoke/fumes become the greatest threat:

Smoke or Fumes Removal Procedure............ Accomplish

See "Smoke or Fumes Removal Procedure" on page 8- 126.

Can the source of smoke/fumes/fire be identified or can the fire be extinguished quickly?

NO → Land immediately at the nearest suitable airport.

CAUTION: DEPENDENT UPON THE SEVERITY OF THE SITUATION, THE FLIGHT CREW SHOULD EXPEDITE THE LANDING. THE CREW SHOULD ALSO CONSIDER AN OVERWEIGHT LANDING, TAILWIND LANDING, OR A FORCED OFF-AIRPORT LANDING.

END

YES ↓

Appropriate Procedure.................................. Accomplish

- Air-conditioning smoke (See "Air-Conditioning System Smoke" on page 8- 129.)
- Electrical smoke or fire (See "Electrical System Smoke" on page 8- 132.)
- Galley smoke or fire (See "Galley Smoke" on page 8- 134.)

CONTINUED ON NEXT PAGE

- Cabin smoke or fire (See “Cabin Smoke” on page 8- 135.)
- Cargo bay smoke (See “SMOKE AFT CARGO or SMOKE FWD CARGO” on page 8- 138.)
- Toilet smoke (See “SMOKE AFT LAV or SMOKE FWD LAV” on page 8- 137.)

Air-Conditioning System Smoke

Is the APU supplying bleed air to the packs?

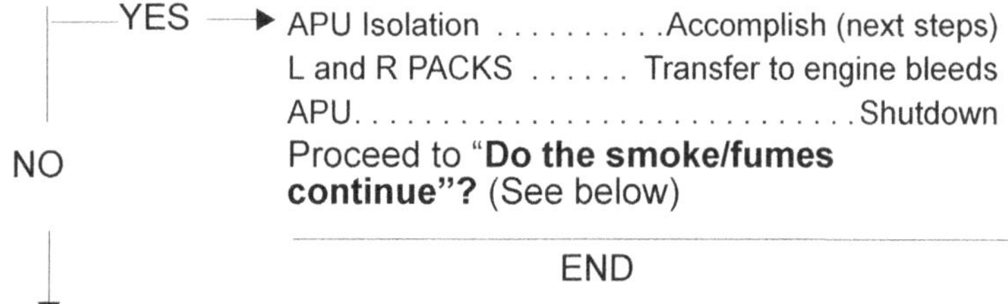

The engines are supplying bleed air to the packs.

Manual Bleed Procedure............ Accomplish (next steps) to select only one engine as the bleed source.

Altitude ....Not above 25000 feet

ISOL switch ....CLSD

BLEED SOURCE selector .... Select alternate source

BLEED VALVES selector ....MANUAL

Affected PACK switch ....OFF

WING A/I CROSS BLEED selector .... Select source engine side

Affected COWL ANTI-ICE switch......Select non-source engine side OFF

Leave icing conditions to prevent ice accumulation on inoperative cowl.

CONTINUED ON NEXT PAGE

NOTE:
Icing conditions exist in-flight at a TAT of 10°C (50°F) or below, and visible moisture in any form is encountered (such as clouds, rain, snow, sleet or ice crystals), except when the SAT is -40°C (-40°F) or below

Proceed to "**Do the smoke/fumes continue**"? (See below).

Do the smoke/fumes continue?

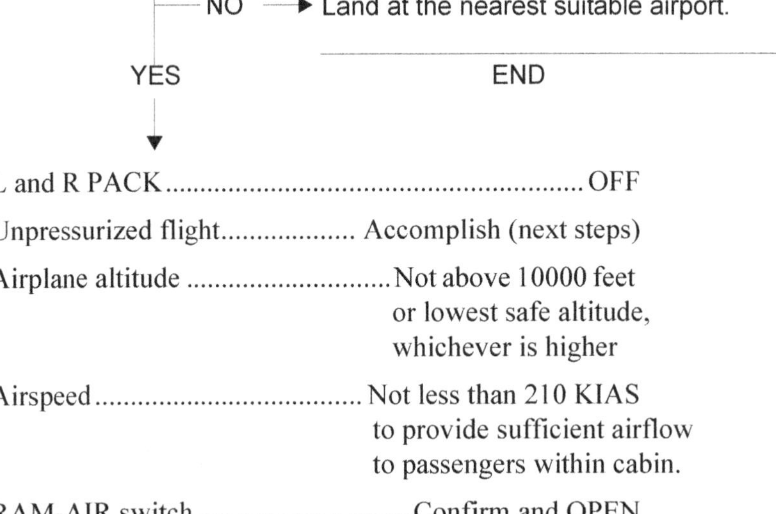

L and R PACK .... OFF

Unpressurized flight.................. Accomplish (next steps)

Airplane altitude .... Not above 10000 feet or lowest safe altitude, whichever is higher

Airspeed .... Not less than 210 KIAS to provide sufficient airflow to passengers within cabin.

RAM-AIR switch.............................. Confirm and OPEN

EMER DEPRESS switch...................... Confirm and ON

Is the source of smoke/fumes/fire visually confirmed to be extinguished or fire controllable?

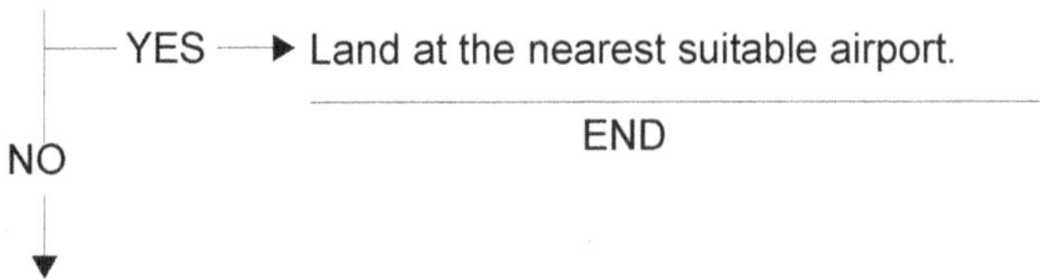

CONTINUED ON NEXT PAGE

Land immediately at the nearest suitable airport.

CAUTION:
Expedite the landing. dependent upon the severity of the situation, the crew should consider an overweight landing, tailwind landing, or ditching or a forced off-airport landing.

Consider passenger evacuation. See "Passenger Evacuation" on page 8-348.

Electrical System Smoke

WARNING

If the affected electrical system cannot be positively identified and isolated, land immediately at the nearest suitable airport.

WARNING

If time and conditions permit, consider determining the source of smoke/fire/fumes by isolating each electrical system one at a time

Affected electrical systems .... Isolate (see table below)

IDENTIFIED BUS	APPLICABLE CIRCUIT BREAKER / SWITCH
AC BUS 1	AC ESS XFER switch . . . . .ALTN Left AUTO XFER switch . . . . .OFF GEN 1 switch. . . . . OFF/RESET
AC BUS 2	Right AUTO XFER switch . . . . .OFF GEN 2 switch. . . . . OFF/RESET
AC ESS BUS	AVIONICS FAN switch . . . . FLT ALTN AC ESS FEED cb(1S2) . . . . Open
AC SERV BUS	AC SERVICE FEED cb(2E2). . . . . Open
DC BUS 1 DC BUS 2	DC 1 FEED cb(1D6) . . . . Open DC 2 FEED cb(2L8). . . . . Open
DC ESS BUS	DC ESS FEED cb (2R6) . . . . Open
DC EMER BUS	APU BATT DIRECT FEED cb (1R1) . . Open EMER BUS FEED cb (1L10). . . . . Open
DC UTILITY	DC UTILITY FEED cb (2L7) . . . . Open
DC SERV BUS	DC SERV BUS FEED cb(2F5) . . . . Open DC SERVICE switch . . . . .OFF
DC BATT BUS	Airplane altitude. . . . . Not above 13000 feet PRESS CONT switch . . . . .Auto LDG ELEV. . . . . .Set to 14000 feet BATT BUS FEED cb (2N2) . . . . Open

CONTINUED ON NEXT PAGE

Affected airplane systems .... Review

See Electrical EICAS message index on page 8-ii.

If at any time the smoke/fumes become the greatest threat, or after the source has been eliminated:

Smoke or Fumes Removal Procedure............ Accomplish

See "Smoke or Fumes Removal Procedure" on page 8- 126.

Is the source of smoke/fumes/fire visually confirmed to be extinguished or fire controllable?

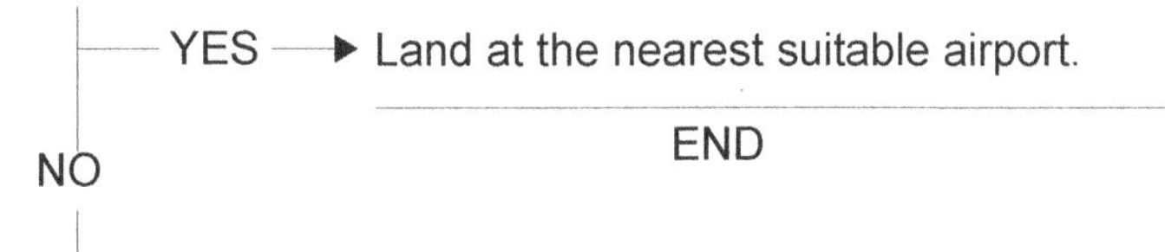

Land immediately at the nearest suitable airport.

CAUTION:
Expedite the landing. dependent upon the severity of the situation, the crew should consider an overweight landing, tailwind landing, or ditching or a forced off-airport landing.

Consider passenger evacuation. See "Passenger Evacuation" on page 8-348.

Galley Smoke

Flight attendant .... Advise to isolate and extinguish source of smoke/fire, secure the area, and open ALL galley control panel circuit breakers.

Affected galley, galley heating system and water system.. Isolate

System	Applicable Circuit Breaker / Switch
Water	WATER SYSTEM cb (2D8). . . . . Open WASTE SYST cb (2M9) . . . . Open WATER CONT cb (2M10). . . . . Open
Galley	GALLEY HEATER CONT cb (2F11) Open GALLEY 1 HEATER cb (2B11) . . . . Open GALLEY EXHAUST FAN cb (2B8). Open
Lights	LIGHTS GALLEY AREA cb (2M6) . Open

If at any time the smoke/fumes become the greatest threat, or after the source has been eliminated:

Smoke or Fumes Removal Procedure............ Accomplish

See "Smoke or Fumes Removal Procedure" on page 8- 126.

Is the source of smoke/fumes/fire visually confirmed to be extinguished or fire controllable?

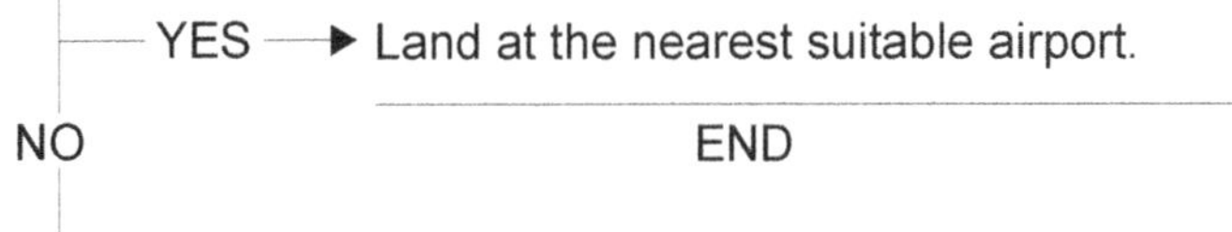

Land immediately at the nearest suitable airport.

CAUTION:
Expedite the landing. dependent upon the severity of the situation, the crew should consider an overweight landing, tailwind landing, or ditching or a forced off-airport landing.

Consider passenger evacuation. See "Passenger Evacuation" on page 8-348.

Cabin Smoke

Flight attendant .... Advise to isolate and extinguish source of smoke/fire and to secure the area.

Affected cabin lighting system .... Isolate

System	Applicable Circuit Breaker / Switch
Cabin Lighting	CABIN CEILING LIGHTS cb (2T2). . . . . Open CABIN LIGHTING cb (2E8). . . . . Open CABIN ACCENT LIGHTS cb (2L4). . . . . Open L CABIN READING LIGHTS cb (1E3). . . . . Open R CABIN READING LIGHTS cb (2L3). . . . . Open LIGHTS CAB UTIL cb (1P4). . . . . Open PASS SIGNS cb (1M10) . . . . Open
Boarding Lights - Door Actuator System	LIGHTS BOARD cb (2M3) . . . . Open PASS DOOR ACT cb (1E1). . . . . Open

If at any time the smoke/fumes become the greatest threat, or after the source has been eliminated:

Smoke or Fumes Removal Procedure............ Accomplish

CONTINUED ON NEXT PAGE

See "Smoke or Fumes Removal Procedure" on page 8- 126.

Is the source of smoke/fumes/fire visually confirmed to be extinguished or fire controllable?

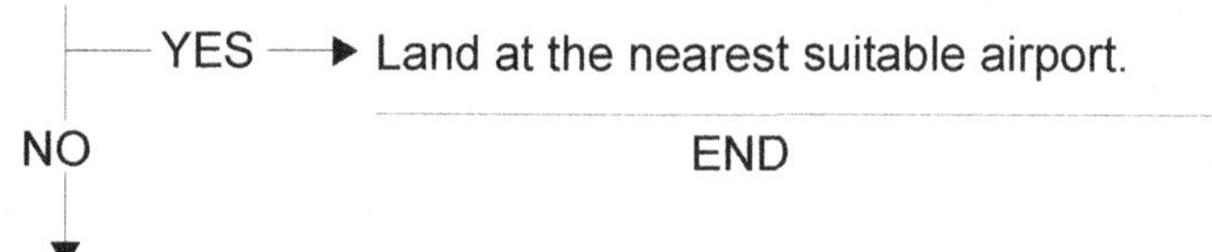

Land immediately at the nearest suitable airport.

CAUTION:

Expedite the landing. dependent upon the severity of the situation, the crew should consider an overweight landing, tailwind landing, or ditching or a forced off-airport landing.

Consider passenger evacuation. See "Passenger Evacuation" on page 8-348.

SMOKE AFT LAV or SMOKE FWD LAV

Flight attendant .... Advise to isolate and extinguish source of smoke/fire and secure the area.

Affected toilet/lavatory area
and water system.. Isolate

System	Applicable Circuit Breaker / Switch
LAV	LAV EXHAUST FAN cb (1B8) . . . . Open
WATER	WATER SYSTEM cb (2D8) . . . . Open WASTE SYST cb (2M9) . . . . Open WATER CONT cb (2M10) . . . . Open
TOILET	TOILET (motor pump) cb(2D5). . . . . Open
LIGHTS	LIGHTS TOILET cb (2M5) . . . . Open

If at any time the smoke/fumes become the greatest threat, or after the source has been eliminated:

Smoke or Fumes Removal Procedure............ Accomplish

See "Smoke or Fumes Removal Procedure" on page 8- 126.

Is the source of smoke/fumes/fire visually confirmed to be extinguished or fire controllable?

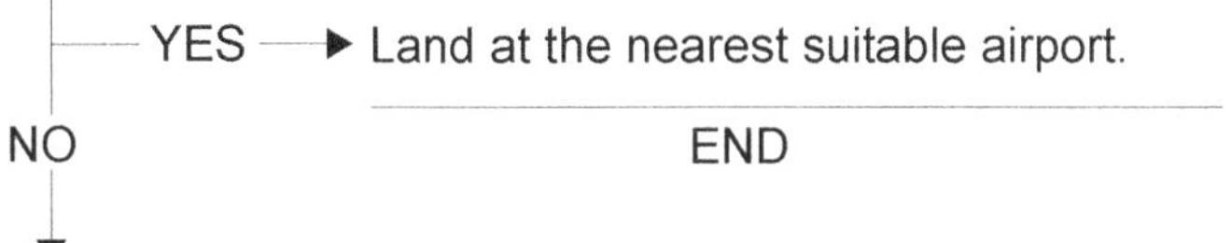

Land immediately at the nearest suitable airport.

CAUTION:

Expedite the landing. dependent upon the severity of the situation, the crew should consider an overweight landing, tailwind landing, or ditching or a forced off-airport landing.

CONTINUED ON NEXT PAGE

Consider passenger evacuation. See "Passenger Evacuation" on page 8-348.

SMOKE AFT CARGO or SMOKE FWD CARGO

Affected CARGO FIREX,
CARGO SMOKE PUSH .... Confirm and Select

Affected FIREX
BOTTLE ARMED-PUSH TO DISCH.................... Select

Squib/bottle discharge... Confirm

LDG ELEV .... Set to 8000 ft. or landing field elevation whichever is higher.

Descent.. Initiate to 10,000 feet MSL or minimum enroute altitude, whichever is higher.

At 10,000 ft. or lowest safe altitude, whichever is higher:

LDG ELEV .... Set to landing field elevation, if not already done.

Land immediately at the nearest suitable airport.

CAUTION:
Expedite the landing. dependent upon the severity of the situation, the crew should consider an overweight landing, tailwind landing, or ditching or a forced off-airport landing.

Consider passenger evacuation. See "Passenger Evacuation" on page 8-348.

CARGO BTL LO

AIR CONDITIONING, AFT CARGO switch .... OFF

Land at the nearest suitable airport.

FWD CARGO DET

Land at the nearest suitable airport.

AFT CARGO DET

Land at the nearest suitable airport.

FIRE SYS FAULT

FIRE DETECTION/FIREX

MONITOR TEST button ....Select

NOTE:
Carry out the appropriate procedure for any or all related messages which come on during the test.

APU SQB

Is APU generator required?

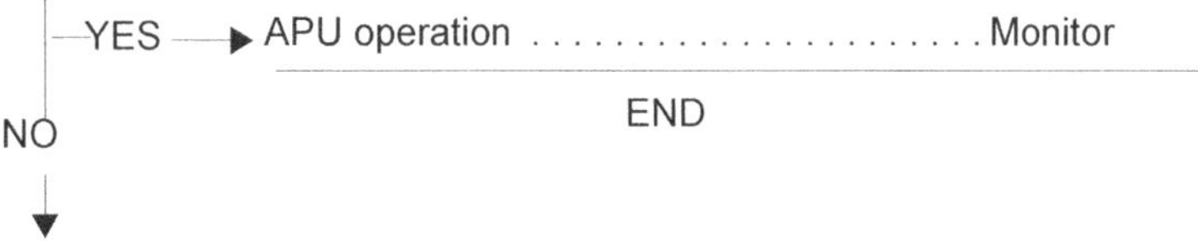

APU START/STOP switch ....Select off

APU PWR FUEL switch ....Select off

L (R) ENG SQB

Fire extinguishing on the affected side is not available.

FWD CARGO SQB 1 (2)

Land at the nearest suitable airport.

AFT CARGO SQB 1 (2)

AIR-CONDITIONING, AFT CARGO switch.......... OFF

Land at the nearest suitable airport.

Hydraulics Index

Intentionally Left Blank

Hydraulics

HYD 1 LO PRESS

NOTE:
If during the accomplishment of a hydraulic system low pressure procedure, a second system also fails, disregard both single system failures and proceed directly to the applicable double system failure procedure.

HYDRAULIC 1 pump..ON

Hydraulic pressure & fluid quantity .... Monitor

CONTINUED ON NEXT PAGE

Is system 1 pressure less than 1800 psi?

YES → HYDRAULIC 1 pump .... OFF

L HYD SOV .... Confirm and CLOSED

Note: The EDP 1A is operating without sufficient hydraulic fluid. Log the length of time that the message remains on display. Monitor hydraulic system 1 fluid temperature readout.

HYDRAULIC and
F/CTL pages .... Review affected systems

HYDRAULIC SYNOPTIC	
COMPONENT	**SYSTEM 1**
L and R Spoilerons (outboard only)	Inoperative
L and R Flight Spoilers (out-board only)	Inoperative
L and R Ground Spoilers (outboard only)	Inoperative
L Thrust Reverser	Inoperative

- OB FLT SPLRS caution msg on:
  - FLIGHT SPOILER lever...............RETRACT
  - Airplane Altitude...................26,000 feet max
- OB GND SPRLRS and OB SPOILERONS caution msgs on:
  - No action required.

Land at the nearest suitable airport.

NO ↓

CONTINUED ON NEXT PAGE

NO

Note: Category II operations may be affected by this failure. Review the equipment requirements in chapter 3 of this manual.

Prior to landing:

GRND PROX, FLAP . . . . OVRD

LH THRUST REVERSER . . . . OFF

Landing flaps . . . . Use 20°

Appch speed . . . . . $V_{REF\ (Flaps\ 45°)}$ + 12 KIAS min.

Note: The landing distance factors below are based upon the loss of the outboard multi-function spoilers, outboard ground spoilers and left thrust reverser.

Actual landing distance . . . .

Without two Thrust Reversers and/or Wet/ Contaminated Runway Surface	With two Thrust Reversers and a Dry Runway Surface
1.45 (45%)	1.40 (40%)

Note: An asymmetric thrust condition will exist using the thrust reverser system with the left thrust reverser not operating. Rudder control assistance (on ground) at high speed will be required to maintain directional control.

END

No further action required.

HYD 2 LO PRESS

NOTE:

If during the accomplishment of a hydraulic system low pressure procedure, a second system also fails, disregard both single system failures and proceed directly to the applicable double system failure procedure.

HYDRAULIC 2 pump..ON

Hydraulic pressure & fluid quantity .... Monitor

CONTINUED ON NEXT PAGE

Is system 2 pressure less than 1800 psi?

YES → HYDRAULIC 2 pump . . . . OFF

R HYD SOV switch . . . . Confirm and CLOSED

Note: The EDP 2A is operating without sufficient hydraulic fluid. Log the length of time that the message remains on display. Monitor hydraulic system 2 fluid temperature readout.

HYDRAULIC and
F/CTL pages . . . . Review affected systems

HYDRAULIC SYNOPTIC	
COMPONENT	**SYSTEM 2**
L and R Spoilerons (inboard only)	Inoperative
L and R Flight Spoilers (inboard only)	Inoperative
Landing Gear	Emergency/alternate extension Inoperative (Downlock assist not available)
R Thrust Reverser	Inoperative
Outboard Brakes	Inoperative once system 2 accumulator pressure is depleted

- IB FLT SPLRS caution msg on:
  - FLIGHT SPOILER lever.................RETRACT
  - Airplane Altitude.................26,000 feet max
- IB SPOILERONS and OB BRAKE PRESS caution msgs on:
  - No action required.
- Landing gear downlock assist not available.

Note: Category II operations may be affected by this failure. Review the equipment requirements in chapter 3 of this manual.

Land at the nearest suitable airport.

NO ↓

CONTINUED ON NEXT PAGE

Prior to landing:

GRND PROX, FLAP . . . . .OVRD

RH THRUST REVERSER . . . . OFF

Landing flaps. . . . . Use 20°

Appch speed . . . . . . $V_{REF\ (Flaps\ 45°)}$ + 12 KIAS min.

NO

Note: The landing distance factors below are based upon the loss of the inboard multi-function spoilers, outboard brakes, and right thrust reverser.

Actual landing distance . . . . Increase.

Without two Thrust Reversers and/or Wet/ Contaminated Runway Surface	With two Thrust Reversers and a Dry Runway Surface
1.90 (90%)	1.75 (75%)

Note: An asymmetric thrust condition will exist using the thrust reverser system with the right thrust reverser not operating. Rudder control assistance (on ground) at high speed will be required to maintain directional control.

CAUTION: ANTICIPATE THE LOSS OF OUTBOARD BRAKES DURING LANDING WHEN SYSTEM 2 BRAKE ACCUMULATOR DEPRESSURIZES.

Brake pressure . . . . .Check

Is brake pressure greater than 1800 psi for the outboard brakes?

NO YES

Use a steady brake application upon landing. Do not cycle brakes.

END

Max landing weight . . . . Determine using the following max brake energy table and correct for wind and slope.

Note: If landing within 25 minutes after takeoff, reduce the maximum landing weight given in the table by 3000 lbs

CONTINUED ON NEXT PAGE

OAT		Airport Pressure Altitude (Feet)					
		0	**2000**	**4000**	**6000**	**8000**	**10000**
°C	°F	**Landing Weight (lbs.) Due to Max Brake Energy**					
-40	-40	84,371	81,053	77,876	74,852	71,894	68,964
-20	-4	80,638	77,553	74,562	71,649	68,803	66,013
0	32	77,300	74,414	71,564	68,767	66,037	63,359
20	68	74,350	71,576	68,861	66,187	63,558	60,981
40	104	71,693	69,015	66,400	63,841	61,290	58,786

NO

Wind corrections:

Increase max landing weight by 2200 lbs per 10 kts headwind.

Decrease max landing weight by 9700 lbs per 10 kts tailwind

Runway Slope Corrections:

Increase max landing weight by 1320 lbs per 1% uphill slope

Decrease max landing weight by 1760 lbs per 1% downhill slope.

Note: The actual landing weight must not exceed the corrected maximum landing weight due to brake energy.

END

No further action required.

HYD 3 LO PRESS

NOTE:

If during the accomplishment of a hydraulic system low pressure procedure, a second system also fails, disregard both single system failures and proceed directly to the applicable double system failure procedure.

HYDRAULIC 3B pump ....ON

Hydraulic pressure & fluid quantity .... Monitor

CONTINUED ON NEXT PAGE

Is system 3 pressure less than 1800 psi?

YES → HYDRAULIC 3A and 3B pumps . . . . OFF

HYDRAULIC and F/CTL pages . . . . Review affected systems

HYDRAULIC SYNOPTIC	
COMPONENT	**SYSTEM 3**
L and R Ground Spoilers (inboard only)	Inoperative
Landing Gear	Normal Extension Inoperative
Nose Wheel Steering	Inoperative
Inboard Brakes	Inoperative once system 3 accumulator pressure is depleted
Parking Brake	Inoperative once system 3 accumulator pressure is depleted.

- If IB GND SPLRS and IB BRAKE PRESS caution messages are on:
  - No action required.

Land at the nearest suitable airport

Note: Category II operations may be affected by this failure. Review the equipment requirements in chapter 3 of this manual.

NO ↓

CONTINUED ON NEXT PAGE

NO

Prior to landing:

GRND PROX, FLAP. . . . . OVRD
N/W STRG switch . . . . OFF
Landing flaps . . . . Use 20°
Appch speed . . . . $V_{REF\ (Flaps\ 45°)}$ + 12 KIAS min.
LDG GEAR lever . . . . DN
LDG GEAR MAN RELEASE . . . . Pull to full extension

Note: The landing distance factors below are based upon the loss of the inboard ground spoilers and inboard brakes.

Actual landing distance . . . . Increase

Without two Thrust Reversers and/or Wet/ Contaminated Runway Surface	With two Thrust Reversers and a Dry Runway Surface
2.00 (100%)	1.70 (70%)

Notes:
1. Select the longest runway available with minimal turbulence and crosswind.
2. Use differential braking, rudder and engine thrust as required to assist directional control.
3. Maximize the use of reverse thrust.

CAUTION: ANTICIPATE THE LOSS OF INBOARD BRAKES DURING LANDING WHEN SYSTEM 3 BRAKE ACCUMULATOR DEPRESSURIZES.

Brake pressure . . . . Check

Is brake pressure greater than 1800 psi for the inboard brakes?

NO YES

YES: Use a steady brake application upon landing. Do not cycle brakes.

END

NO: Max landing weight . . . . Determine using the following max brake energy table and correct for wind and slope.

CONTINUED ON NEXT PAGE

NO

Note: If landing within 25 minutes after takeoff, reduce the maximum landing weight given in the table by 3000 lbs.

OAT		Airport Pressure Altitude (Feet)					
		0	**2000**	**4000**	**6000**	**8000**	**10000**
°C	**°F**	**Landing Weight (lbs.) Due to Max Brake Energy**					
-40	-40	84,371	81,053	77,876	74,852	71,894	68,964
-20	-4	80,638	77,553	74,562	71,649	68,803	66,013
0	32	77,300	74,414	71,564	68,767	66,037	63,359
20	68	74,350	71,576	68,861	66,187	63,558	60,981
40	104	71,693	69,015	66,400	63,841	61,290	58,786

Wind corrections:

Increase max landing weight by 2200 lbs per 10 kts headwind.

Decrease max landing weight by 9700 lbs per 10 kts tailwind

Runway Slope Corrections:

Increase max landing weight by 1320 lbs per 1% uphill slope

Decrease max landing weight by 1760 lbs per 1% downhill slope.

Note: The actual landing weight must not exceed the corrected maximum landing weight due to brake energy.

END

No further action required.

HYD 1 HI TEMP

Indication: HYD 1 HI TEMP caution message on and temperature readouts greater than 96°C (205°F).

NOTE:
If the hydraulic temperature increases to between 120°C (245°F) and 140°C (280°F) a hydraulic fuse will open and drain the contents of the hydraulic system.

L HYD SOV switch......................Confirm and CLOSED

HYDRAULIC page ....Confirm L HYD SOV closed

System 1 temperature .... Monitor

CONTINUED ON NEXT PAGE

Is the L HYD SOV closed OR System 1 temperature stabilized?

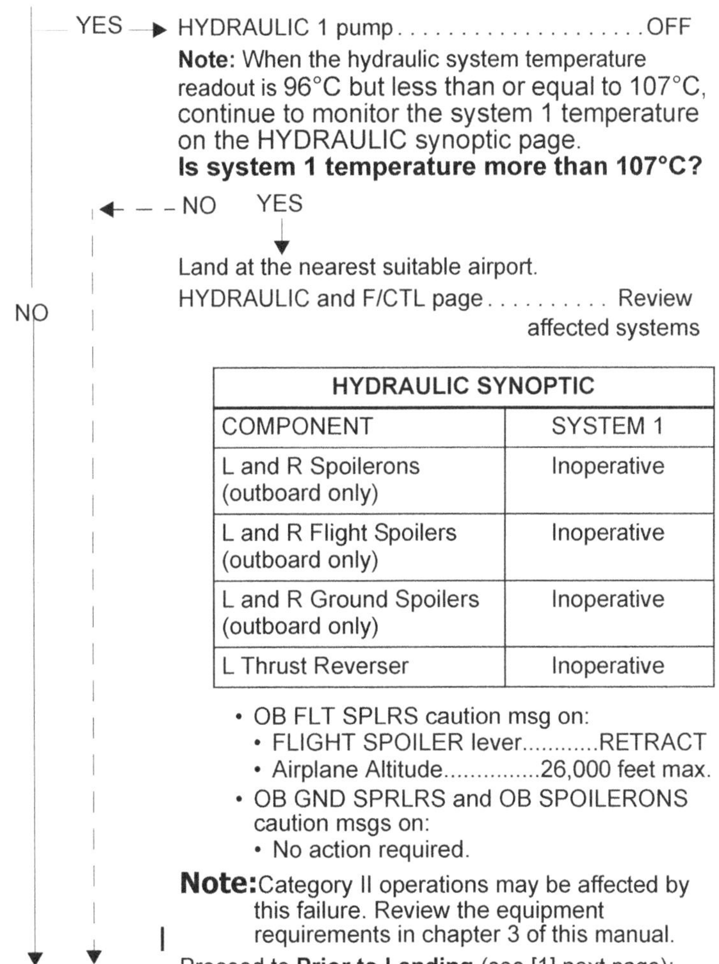

YES → HYDRAULIC 1 pump .OFF

Note: When the hydraulic system temperature readout is 96°C but less than or equal to 107°C, continue to monitor the system 1 temperature on the HYDRAULIC synoptic page.

Is system 1 temperature more than 107°C?

← NO YES ↓

Land at the nearest suitable airport.

HYDRAULIC and F/CTL page . . . . Review affected systems

HYDRAULIC SYNOPTIC	
COMPONENT	SYSTEM 1
L and R Spoilerons (outboard only)	Inoperative
L and R Flight Spoilers (outboard only)	Inoperative
L and R Ground Spoilers (outboard only)	Inoperative
L Thrust Reverser	Inoperative

- OB FLT SPLRS caution msg on:
  - FLIGHT SPOILER lever............RETRACT
  - Airplane Altitude...............26,000 feet max.
- OB GND SPRLRS and OB SPOILERONS caution msgs on:
  - No action required.

Note: Category II operations may be affected by this failure. Review the equipment requirements in chapter 3 of this manual.

NO ↓

Proceed to **Prior to Landing** (see [1] next page):

CONTINUED ON NEXT PAGE

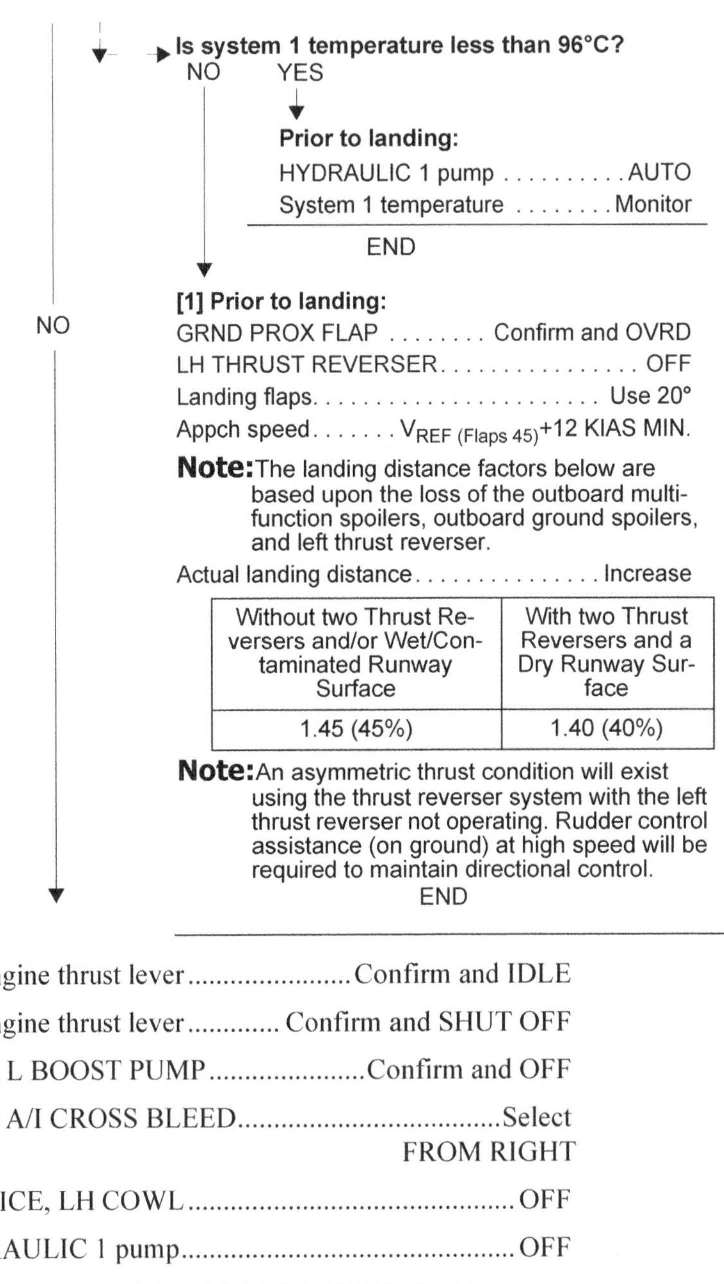

Is system 1 temperature less than 96°C?

NO YES

Prior to landing:

HYDRAULIC 1 pump . . . . AUTO

System 1 temperature . . . . Monitor

END

[1] Prior to landing:

NO

GRND PROX FLAP . . . . Confirm and OVRD

LH THRUST REVERSER. . . . . OFF

Landing flaps. . . . . Use 20°

Appch speed. . . . . . . $V_{REF\ (Flaps\ 45)}$+12 KIAS MIN.

Note: The landing distance factors below are based upon the loss of the outboard multi-function spoilers, outboard ground spoilers, and left thrust reverser.

Actual landing distance. . . . . Increase

Without two Thrust Reversers and/or Wet/Contaminated Runway Surface	With two Thrust Reversers and a Dry Runway Surface
1.45 (45%)	1.40 (40%)

Note: An asymmetric thrust condition will exist using the thrust reverser system with the left thrust reverser not operating. Rudder control assistance (on ground) at high speed will be required to maintain directional control.

END

Left engine thrust lever........................Confirm and IDLE

Left engine thrust lever............. Confirm and SHUT OFF

FUEL, L BOOST PUMP......................Confirm and OFF

WING A/I CROSS BLEED....................................Select FROM RIGHT

ANTI-ICE, LH COWL..OFF

HYDRAULIC 1 pump..OFF

CONTINUED ON NEXT PAGE

APU (if available) (37,000 feet and below)..........START

Fuel System .... Check

Fuel Crossflow.. AUTO

Fuel Quantity/Balance .... Check

NOTE:
Leave icing conditions to prevent ice accumulation on the engine cowl with the inoperative anti-icing system.

Airspeed...................................Not more than 280 KIAS

Land at the nearest suitable airport.

HYDRAULIC and F/CTL page............. Review affected systems

Hydraulic Synoptic	
COMPONENT	**SYSTEM 1**
L and R Spoilerons (outboard only	Inoperative
L and R Flight Spoilers (outboard only)	Inoperative
L and R Ground Spoilers (outboard only)	Inoperative
L Thrust Reverser	Inoperative

OB FLT SPLRS caution msg on:

- FLIGHT SPOILER lever............................RETRACT
- Airplane Altitude..................................26,000 feet max

OB GND SPRLRS and OB SPOILERONS caution msgs on:

- No action required.

NOTE:
Category II operations may be affected by this failure. Review the equipment requirements in chapter 3 of this manual.

CONTINUED ON NEXT PAGE

Prior to landing:

GRND PROX, FLAP ....Confirm and OVRD

LH THRUST REVERSER ....OFF

Landing flaps ....Use 20°

Approach Speed.. $V_{REF\ (Flaps\ 45°)}$ + 12 KIAS MINIMUM

NOTE:
The landing distance factors below are based upon the loss of the outboard multi-function spoilers, outboard ground spoilers, left thrust reverser, and left engine inoperative.

Actual landing distance....................................... Increase.

Without two Thrust Reversers and/or Wet/ Contaminated Runway Surface	With two Thrust Reversers and a Dry Runway Surface
1.45 (45%)	1.40 (40%)

NOTE:
An asymmetric thrust condition will exist using the thrust reverser system with the left thrust reverser not operating. Rudder control assistance (on ground) at high speed will be required to maintain directional control.

HYD 2 HI TEMP

NOTE:
If the hydraulic temperature increases to between 120°C (245°F) and 140°C (280°F) a hydraulic fuse will open and drain the contents of the hydraulic system.

R HYD SOV switch ....Confirm and CLOSED

HYDRAULIC page............Confirm R HYD SOV closed

System 2 temperature ....Monitor

CONTINUED ON NEXT PAGE

Is the R HYD SOV closed OR System 2 temperature stabilized?

YES → HYDRAULIC 2 pump . . . . OFF

Note: When the hydraulic system temperature readout is 96°C but less than or equal to 107°C, continue to monitor the system 2 temperature on the HYDRAULIC synoptic page.

Is system 2 temperature more than 107°C?

NO YES

Land at the nearest suitable airport.

HYDRAULIC and F/CTL page . . . . Review affected systems

NO

HYDRAULIC SYNOPTIC	
COMPONENT	SYSTEM 2
L and R Spoilerons (inboard only)	Inoperative
L and R Flight Spoilers (inboard only)	Inoperative
Landing Gear	Emergency/alternate extension Inoperative (downlock assist not available)
R Thrust Reverser	INOPERATIVE
Outboard Brakes	Inoperative once system 2 accumulator pressure is depleted.

NO

- IB FLT SPLRS caution msg on:
  - FLIGHT SPOILER lever.............RETRACT
  - Airplane Altitude................26,000 feet max.
- IB SPOILERONS and OB BRAKE PRESSURE caution msgs on:
  - No action required.

Note: Category II operations may be affected by this failure. Review the equipment requirements in chapter 3 of this manual.

Proceed to **Prior to Landing** (see [1] next page):

CONTINUED ON NEXT PAGE

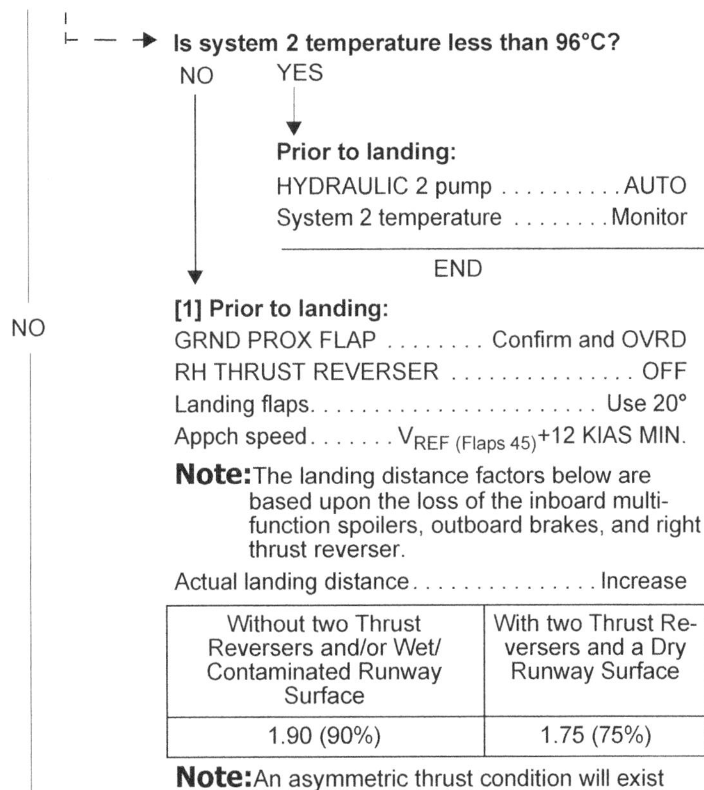

Is system 2 temperature less than 96°C?

NO YES

Prior to landing:

HYDRAULIC 2 pump . . . . AUTO
System 2 temperature . . . . Monitor

END

NO

[1] Prior to landing:

GRND PROX FLAP . . . . Confirm and OVRD
RH THRUST REVERSER . . . . OFF
Landing flaps. . . . . Use 20°
Appch speed. . . . . . . $V_{REF\ (Flaps\ 45)}$+12 KIAS MIN.

Note: The landing distance factors below are based upon the loss of the inboard multi-function spoilers, outboard brakes, and right thrust reverser.

Actual landing distance. . . . . Increase

Without two Thrust Reversers and/or Wet/ Contaminated Runway Surface	With two Thrust Reversers and a Dry Runway Surface
1.90 (90%)	1.75 (75%)

Note: An asymmetric thrust condition will exist using the thrust reverser system with the right thrust reverser not operating. Rudder control assistance (on ground) at high speed will be required to maintain directional control.

CAUTION: ANTICIPATE THE LOSS OF OUTBOARD BRAKES DURING LANDING WHEN THE SYSTEM 2 BRAKE ACCUMULATOR

CONTINUED ON NEXT PAGE

Brake pressure . . . . .Check

Is brake pressure greater than 1800 psi for the outboard brakes?

NO YES

YES: Use a steady brake application upon landing. Do not cycle brakes.

END

NO

Max landing weight. . . . . Determine using the following max brake energy table and correct for wind and slope.

Note: If landing within 25 minutes after takeoff, reduce the maximum landing weight given in the table by 3000 lbs.

OAT		Airport Pressure Altitude (Feet)					
		0	**2000**	**4000**	**6000**	**8000**	**10000**
°C	**°F**	**Landing Weight (lbs.) Due to Max Brake Energy**					
-40	-40	84,371	81,053	77,876	74,852	71,894	68,964
-20	-4	80,638	77,553	74,562	71,649	68,803	66,013
0	32	77,300	74,414	71,564	68,767	66,037	63,359
20	68	74,350	71,576	68,861	66,187	63,558	60,981
40	104	71,693	69,015	66,400	63,841	61,290	58,786

Wind corrections:

Increase max landing weight by 2200 lbs per 10 kts headwind.

Decrease max landing weight by 9700 lbs per 10 kts tailwind

Runway Slope Corrections:

Increase max landing weight by 1320 lbs per 1% uphill slope

Decrease max landing weight by 1760 lbs per 1% downhill slope.

Note: The actual landing weight must not exceed the corrected maximum landing weight due to brake energy.

END

CONTINUED ON NEXT PAGE

Right engine thrust lever....................Confirm and IDLE

Right engine thrust lever.......... Confirm and SHUT OFF

FUEL, R BOOST PUMP.....................Confirm and OFF

WING A/I CROSS BLEED....................................Select FROM LEFT

ANTI-ICE, RH COWL...OFF

HYDRAULIC 2 pump..OFF

APU (if available) (37,000 feet and below)..........START

Fuel System ....Check

Fuel Crossflow...AUTO

Fuel Quantity/Balance ....Check

NOTE:
Leave icing conditions to prevent ice accumulation on the engine cowl with the inoperative anti-icing system.

Airspeed...................................Not more than 280 KIAS

Land at the nearest suitable airport.

HYDRAULIC and F/CTL page............................ Review affected systems

Hydraulic Synoptic	
COMPONENT	**SYSTEM 2**
L and R Spoilerons (inboard only	Inoperative
L and R Flight Spoilers (inboard only)	Inoperative
Landing gear	Emergency/alternate extension Inoperative (downlock assist not available)
R Thrust Reverser	Inoperative
Outboard brakes	Inoperative once system 2 accumulator pressure is depleted.

CONTINUED ON NEXT PAGE

IB FLT SPLRS caution msg on:

- FLIGHT SPOILER lever......................RETRACT
- Airplane Altitude........................26,000 FEet max

IB SPOILERONS and OB BRAKE PRESS caution msgs on:

- No action required.

NOTE:
Category II operations may be affected by this failure. Review the equipment requirements in chapter 3 of this manual.

Prior to landing:

GRND PROX, FLAP ....Confirm and OVRD

RH THRUST REVERSER..OFF

Landing flaps ....Use 20°

Approach Speed......$V_{REF\ (Flaps\ 45°)}$ + 12 KIAS minimum

NOTE:
The landing distance factors below are based upon the loss of the inboard multi-function spoilers, outboard brakes, right thrust reverser, and right engine inoperative.

Actual landing distance....................................... Increase.

Without two Thrust Reversers and/or Wet/ Contaminated Runway Surface	With two Thrust Reversers and a Dry Runway Surface
1.90 (90%)	1.75 (75%)

CONTINUED ON NEXT PAGE

NOTE:
An asymmetric thrust condition will exist using the thrust reverser system with the right thrust reverser not operating. Rudder control assistance (on ground) at high speed will be required to maintain directional control.

CAUTION:
Anticipate the loss of outboard brakes during landing when the system 2 brake accumulator depressurizes.

Brake pressure .... Check

HYD 3 HI TEMP

NOTE:
If the HYDRAULIC 3A and 3B pumps were running at the time the HYD 3 HI TEMP caution message appeared, attempt to isolate the malfunctioning pump by switching each pump to OFF, in turn.

HYDRAULIC pump 3B ....ON

HYDRAULIC pump 3A ....OFF

System 3 temperature.. Monitor

Does the temperature stabilize or decrease?

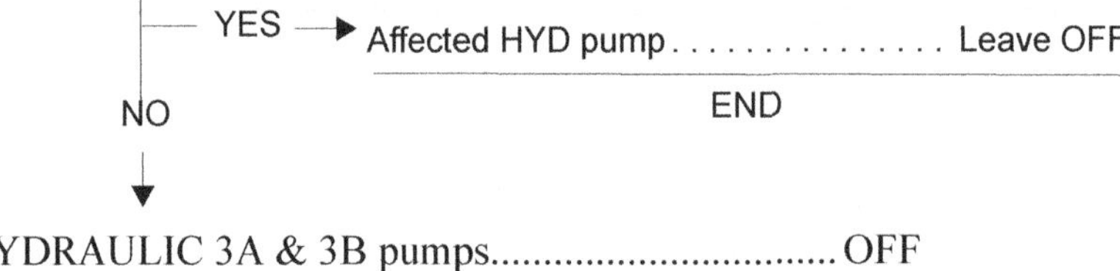

HYDRAULIC 3A & 3B pumps...............................OFF

System 3 temperature.. Monitor

NOTE:
No action is required for the HYD 3 LO PRESS caution message, continue HYD 3 HI TEMP procedure until landing.

NOTE:
Category II operations may be affected by this failure. review the equipment requirements found in chapter 3 of this manual.

Is the temperature less than 107°C?

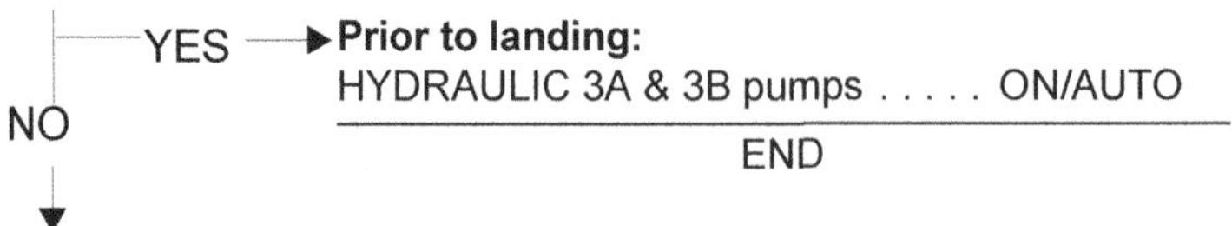

Land at the nearest suitable airport.

Prior to landing:

HYDRAULIC 3A & 3B pumps..................... ON/AUTO

HYD EDP 1A (2A)

Associated HYDRAULIC (1 or 2) pump ....ON

NOTE:
The applicable EDP may be operating without sufficient hydraulic fluid. Log the length of time that the message remains on display. Monitor the applicable hydraulic system fluid temperature readout.

Hydraulic pressure & fluid quantity .... Monitor

NOTE:
Category II operations may be affected by this failure. Review the equipment requirements found in chapter 3 of this manual.

HYD PUMP 1B

HYDRAULIC 1 pump...ON

Hydraulic pressure & fluid quantity .... Monitor

Does the HYD PUMP 1B caution message persist?

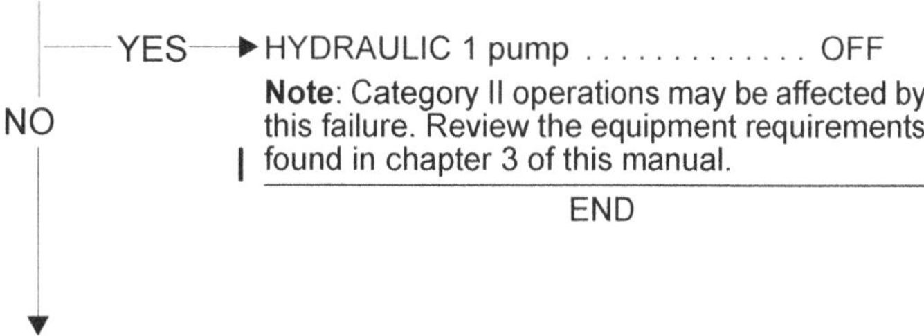

No further action required.

HYD PUMP 2B

HYDRAULIC 2 pump..ON

Hydraulic pressure & fluid quantity .... Monitor

Does the HYD PUMP 2B caution message persist?

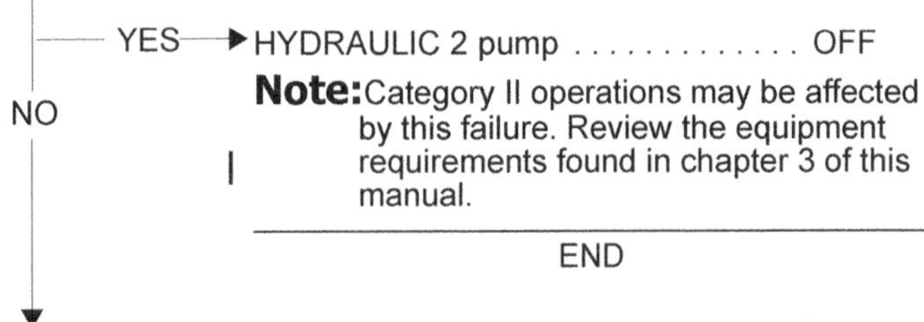

No further action required.

HYD PUMP 3A

HYDRAULIC 3B pump..ON

HYDRAULIC 3A pump..OFF

Hydraulic pressure and fluid quantity.................. Monitor

NOTE:

Category II operations may be affected by this failure. Review the equipment requirements found in chapter 3 of this manual.

HYD PUMP 3B

HYDRAULIC 3B pump ....ON

Hydraulic pressure and fluid quantity.................. Monitor

Does the HYD PUMP 3B caution message persist?

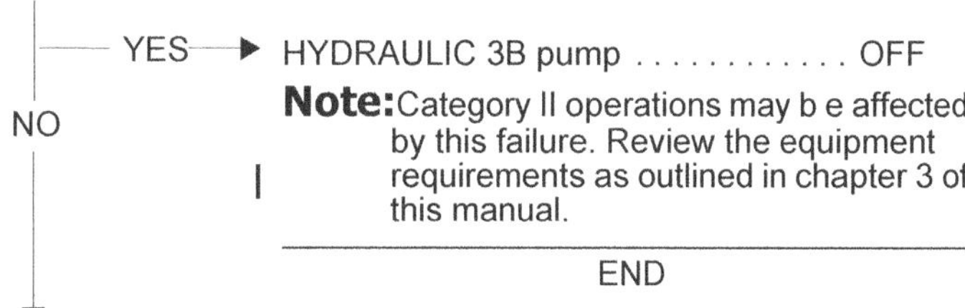

No further action required.

HYD SOV 1 (2) OPEN

Land at the nearest suitable airport.

NOTE:
Category II operations may be affected by this failure. Review the equipment requirements as outlined in chapter 3 of this manual.

HYD 1 LO PRESS and HYD 2 LO PRESS

HYDRAULIC pumps (all) ....ON

Systems 1 and 2 pressure and fluid quantity .... Check

Are systems 1 & 2 pressures & fluid quantities normal?

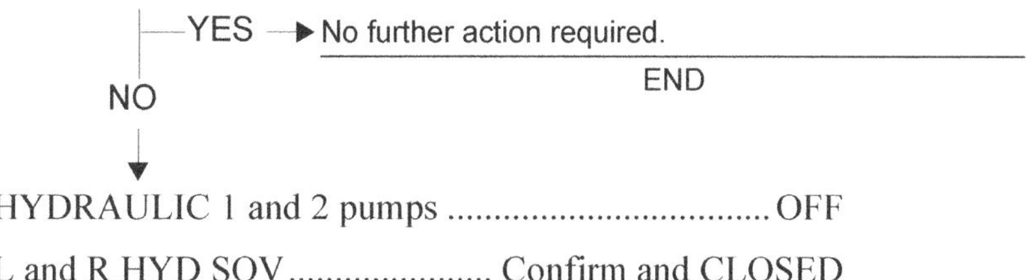

HYDRAULIC 1 and 2 pumps .... OFF

L and R HYD SOV...................... Confirm and CLOSED

CONTINUED ON NEXT PAGE

NOTE:
The engine driven pumps (EDPs) 1A and 2A are operating without sufficient hydraulic fluid. Log the length of time that the messages remain on display. Monitor hydraulic system 1 and 2 fluid temperature readout.

HYDRAULIC page and
F/CTL page .... Review affected systems

Hydraulic Synoptic	
COMPONENT	**SYSTEM 1 AND 2**
L and R Flight Spoilers (inboard and outboard)	Inoperative
L and R Spoilerons (inboard and outboard)	Inoperative
L and R Ground Spoilers (outboard only)	Inoperative
Landing gear	Emergency/alternate extension Inoperative (downlock assist not available)
L and R Thrust Reversers	Inoperative
Outboard brakes	Inoperative once system 2 accumulator pressure is depleted.

IB FLT SPLR and OB FLT SPLRS caution msg on:

- FLIGHT SPOILER lever..........................RETRACT
- Airplane Altitude..............................26,000 feet max

OB GND SPLRS, IB SPOILERONS, OB SPOILERONS, and OB BRAKE PRESS caution messages on:

- No action required.

NOTE:
Category II operations may be affected by this failure. Review the equipment requirements found in chapter 3 of this manual.

Land at the nearest suitable airport.

CONTINUED ON NEXT PAGE

Prior to landing:

GRND PROX, FLAP........................Confirm and OVRD

LH and RH THRUST REVERSERs.........................OFF

Landing flaps ....Use 20°

Approach speed ....$V_{REF\ (Flaps\ 45)}$ + 12 KIAS minimum

NOTE:
The landing distance factors below are based upon the loss of the inboard/outboard multi-function spoilers, outboard ground spoilers, outboard brakes, and both thrust reversers.

Actual landing distance....................................... Increase

Condition	Without Thrust Reversers
Hydraulic Systems 1 and 2 Failed	2.15 (115%)

CAUTION:
Anticipate the loss of outboard brakes during landing when system 2 brake accumulator depressurizes.

CAUTION:
A steady brake application is recommended upon landing. do not cycle the brakes.

Max landing weight .... Determine using the following max brake energy table and correct for wind and slope.

NOTE:
If landing within 25 minutes after takeoff, reduce the maximum landing weight given in the table by 3000 lbs.

CONTINUED ON NEXT PAGE

OAT		Airport Pressure Altitude (Feet)					
		0	2000	4000	6000	8000	10000
°C	°F	**Landing Weight (lbs.) Due to Max Brake Energy**					
-40	-40	83,318	80,055	76,924	73,946	71,027	68,130
-20	-4	79,617	76,600	73,650	70,773	67,963	65,206
0	32	76,320	73,473	70,677	67,914	65,219	62,574
20	68	73,393	70,657	67,979	65,357	62,760	60,216
40	104	70,758	68,117	65,538	63,014	60,510	58,039

Wind corrections:

Increase max landing weight by 2200 lbs per 10 kts headwind.

Decrease max landing weight by 9700 lbs per 10 kts tailwind

Runway Slope Corrections:

Increase max landing weight by 1540 lbs per 1% uphill slope

Decrease max landing weight by 2200 lbs per 1% downhill slope.

NOTE:
The actual landing weight must not exceed the corrected maximum landing weight due to brake energy.

HYD 2 LO PRESS and HYD 3 LO PRESS

HYDRAULIC pumps (all)..ON

Systems 2 and 3 pressure and fluid qty.................. Check

Are system 2 & 3 pressures and fluid quantity normal?

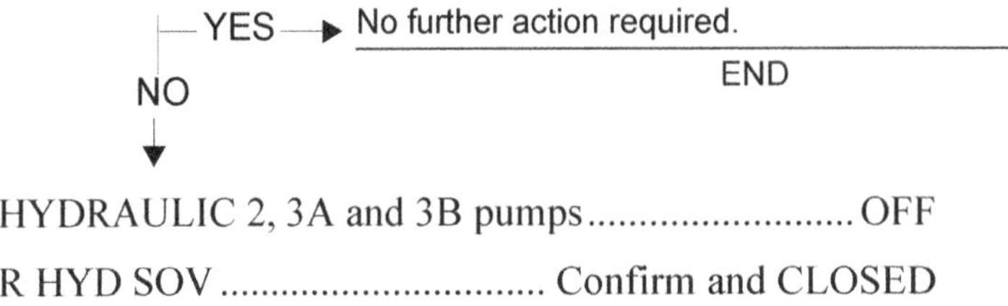

HYDRAULIC 2, 3A and 3B pumps.......................... OFF

R HYD SOV .... Confirm and CLOSED

CONTINUED ON NEXT PAGE

NOTE:
The engine driven pumps (EDPs) 2A is operating without sufficient hydraulic fluid. Log the length of time that the messages remain on display. Monitor hydraulic system 2 fluid temperature readout.

HYDRAULIC and F/CTL page............. Review affected systems

Hydraulic Synoptic	
COMPONENT	**SYSTEM 2 and 3**
R aileron	Inoperative and will upfloat. Use aileron trim as required to compensate.
L and R spoilerons (inboard only)	Inoperative
L and R Flight Spoilers (inboard only)	Inoperative
L and R Ground Spoilers (inboard only)	Inoperative
Landing gear	Normal extension and retraction Inoperative. (MLG may not lock down during manual extension. The nose gear will not extend during the landing gear manual extension procedure).
N/W steering	Inoperative
R Thrust Reverser	Inoperative
Inboard and Outboard Brakes	Inoperative once pressure in system 2 and 3 accumulators is depleted.
Parking Brake	Inoperative once system 3 accumulator pressure is depleted.

IB FLT SPLRS caution msg. on:

- FLIGHT SPOILER lever............................RETRACT
- Airplane Altitude................................26,000 feet max

CONTINUED ON NEXT PAGE

IB GND SPLRS, IB SPOILERONS, IB BRAKE PRESS and OB BRAKE PRESS caution messages on:

- No action required.

NOTE:
Category II operations may be affected by this failure. Review the equipment requirements found in chapter 3 of this manual.

Land at the nearest suitable airport.

Prior to landing:

GRND PROX, FLAP........................Confirm and OVRD

R THRUST REVERSER...OFF

N/W STRG ....OFF

ANTI SKID...OFF

Landing flaps ....Use 20°

LDG GEAR lever ....DN

LDG GEAR MANUAL RELEASE ....Pull to full extension

Approach speed... $V_{REF\,(Flaps\,45)}$ + 12 KIAS MINIMUM

WARNING

The nosegear will not extend during the landing gear manual extension procedure with both "HYD 2 lo press" and "HYD 3 lo press" messages displayed.

Landing gear up/

Unsafe Landing Procedure............................. Accomplish

See "Landing Gear Up / Unsafe Landing Procedure" on page 8- 289.

NOTE:
Flight path control is limited with hydraulic systems 2 and 3 failed.

CONTINUED ON NEXT PAGE

NOTE:
Select the longest runway available with minimal crosswind and turbulence.

NOTE:
Rudder control is adequate for normal flight and should be used in coordination with aileron, if necessary, during turns.

NOTE:
Unless it is possible to restore the use of hydraulic system 2, main landing gear extension relies upon free fall following a manual landing gear extension. Side slip may be required to achieve down lock.

NOTE:
An asymmetric thrust condition will exist, using the thrust reverse system with the right thrust reverser not operating. Rudder control assistance (on ground) at high speed will be required to maintain directional control.

CAUTION:
Anticipate the loss of inboard and outboard brakes when systems 2 and 3 brake accumulators depressurize.

CAUTION:
A steady brake application is recommended upon landing. do not cycle the brakes.

HYD 1 LO PRESS and HYD 3 LO PRESS

HYDRAULIC pumps (all)..ON

Systems 1 and 3 pressure and fluid quantity........... Check

Are systems 1 & 3 pressures & fluid quantities normal?

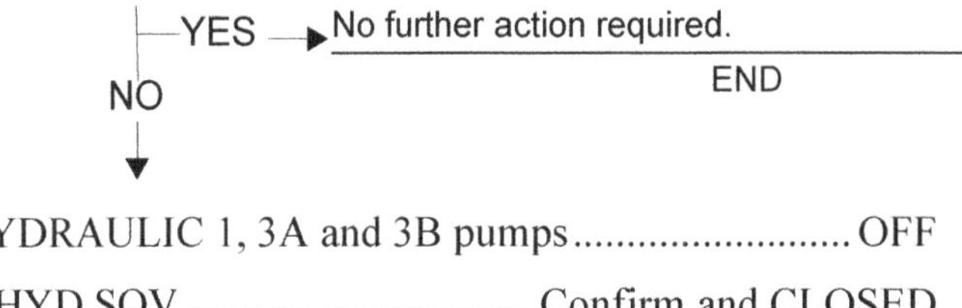

HYDRAULIC 1, 3A and 3B pumps..........................OFF

L HYD SOV .... Confirm and CLOSED

NOTE:
The engine driven pumps (EDP) 1A is operating without sufficient hydraulic fluid. Log the length of time that the messages remain on display. Monitor hydraulic system 1 fluid temperature readout.

HYDRAULIC and F/CTL page............................ Review affected systems

Hydraulic Synoptic	
COMPONENT	**SYSTEM 1 and 3**
L Aileron	Inoperative and will upfloat. Use aileron trim as required to compensate.
L and R Spoilerons (outboard only)	Inoperative
L and R Flight Spoilers (outboard only)	Inoperative
L and R Ground Spoilers (inboard and outboard)	Inoperative
L Thrust Reverser	Inoperative
Landing Gear	Normal extension Inoperative
N/W Steering	Inoperative
Inboard Brakes	Inoperative once system 3 accumulator pressure is depleted.
Parking brake	Inoperative once system 3 accumulator pressure is depleted.

CONTINUED ON NEXT PAGE

OB FLT SPLRS caution message on:

- FLIGHT SPOILER lever....................RETRACT
- Airplane Altitude........................26,000 feet max

IB GND SPLRS, OB GND SPLRS, OB SPOILERONS, and IB BRAKE PRESS caution messages on:

- No action required

NOTE:
Category II operations may be affected by this failure. Review the equipment requirements found in chapter 3 of this manual.

Land at the nearest suitable airport.

Prior to landing:

GRND PROX, FLAP........................Confirm and OVRD

LH THRUST REVERSER .... OFF

N/W STRG .... OFF

Landing flaps ....Use 20°

LDG GEAR lever ....DN

LDG GEAR MANUAL RELEASE .... Pull to full extension

Approach speed ....$V_{REF\ (Flaps\ 45)}$ + 12 KIAS minimum

NOTE:
The landing distance factors below are based upon the loss of the outboard multi-function spoilers, inboard/outboard ground spoilers, inboard brakes, and left thrust reverser.

Actual landing distance....................................... Increase

Without Thrust Reverser	With One Thrust Reverser
2.15 (115%)	1.90 (90%)

CONTINUED ON NEXT PAGE

NOTE:
Flight path control is limited with hydraulic systems 1 and 3 failed.

NOTE:
Select the longest runway available with minimal crosswind and turbulence.

NOTE:
Rudder control is adequate for normal flight and should be used in coordination with aileron, if necessary, during turns.

NOTE:
An asymmetric thrust condition will exist, using the thrust reverse system with the left thrust reverser not operating. Rudder control assistance (on ground) at high speed will be required to maintain directional control.

CAUTION:
Anticipate the loss of inboard brakes upon landing, when system 3 brake accumulator depressurizes.

CAUTION:
A steady brake application is recommended upon landing. do not cycle the brakes.

Max landing weight .... Determine using the following max brake energy table and correct for wind and slope.

NOTE:
If landing within 25 minutes after takeoff, reduce the maximum landing weight given in the table by 3000 lbs.

CONTINUED ON NEXT PAGE

OAT		Airport Pressure Altitude (Feet)					
		0	**2000**	**4000**	**6000**	**8000**	**10000**
°C	**°F**	**Landing Weight (lbs.) Due to Max Brake Energy**					
-40	-40	83,318	80,055	76,924	73,946	71,027	68,130
-20	-4	79,617	76,600	73,650	70,773	67,963	65,206
0	32	76,320	73,473	70,677	67,914	65,219	62,574
20	68	73,393	70,657	67,979	65,357	62,760	60,216
40	104	70,758	68,117	65,538	63,014	60,510	58,039

Wind corrections:

Increase max landing weight by 2200 lbs per 10 kts headwind.

Decrease max landing weight by 9700 lbs per 10 kts tailwind

Runway Slope Corrections:

Increase max landing weight by 1540 lbs per 1% uphill slope

Decrease max landing weight by 2200 lbs per 1% downhill slope.

NOTE:
The actual landing weight must not exceed the corrected maximum landing weight due to brake energy.

Intentionally Left Blank

Fuel Index

Intentionally Left Blank

Fuel

LO FUEL

Is the quantity in either main tank less than 600 lbs. or total fuel less than 1,200 lbs.?

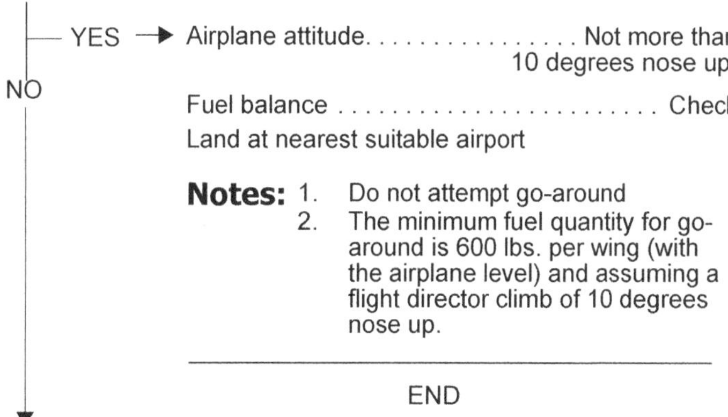

NOTE:
In this condition, both fuel collector cells are low.

Do not climb.

Fuel balance .... Check

Land at the nearest suitable airport.

FUEL CH 1/2 FAIL

Engine thrust .... Adjust as required to maintain equal fuel flow to the engines

Fuel balance .... Maintain using fuel used indication

Land at nearest suitable airport.

Gravity Crossfeed Procedure .... Accomplish

See "Gravity Crossfeed Procedure" on page 8- 190.

L (R) MAIN EJECTOR and L (R) SCAV EJECTOR

FUEL, L and R BOOST PUMPS ....Confirm operating

Fuel tank quantities.. Monitor

Is the total fuel quantity depleting abnormally?

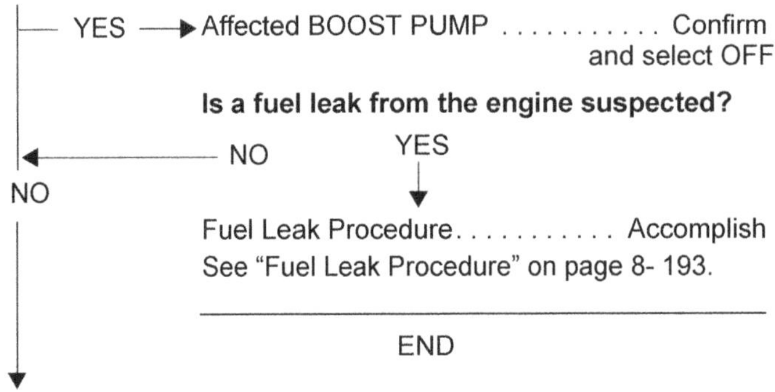

No further action required.

FUEL IMBALANCE

Automatic crossflow ....Confirm operating

Is a leak into the center tank suspected?

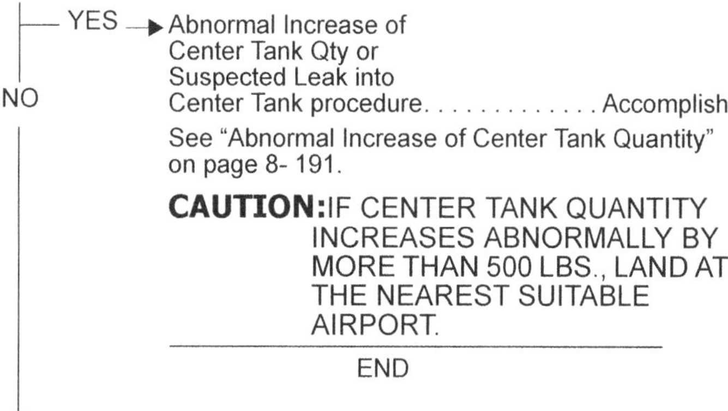

Is automatic fuel crossflow operating and does fuel imbalance persist?

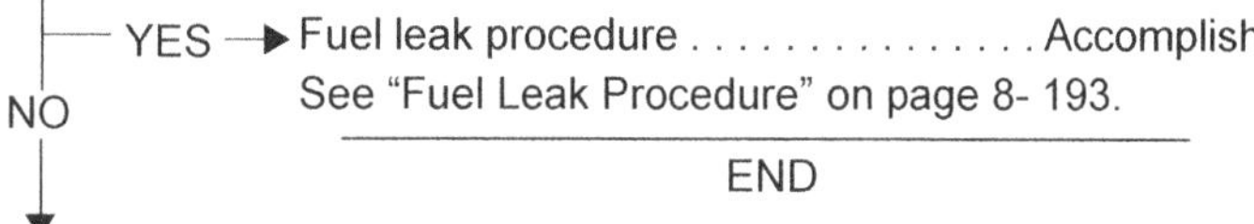

Is automatic crossflow inoperative?

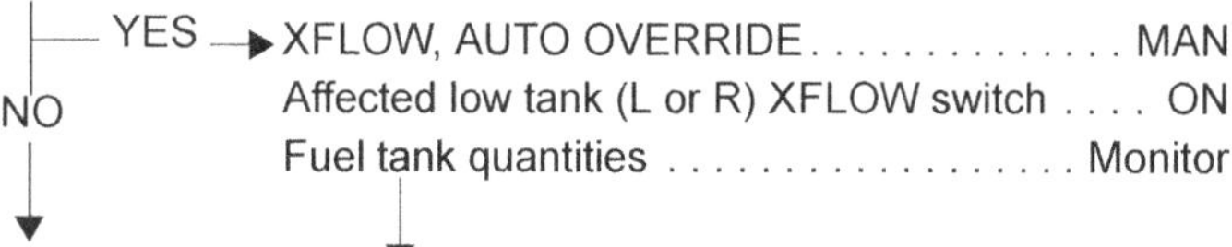

Is a leak from the wing tank suspected (fuel imbalance persists?

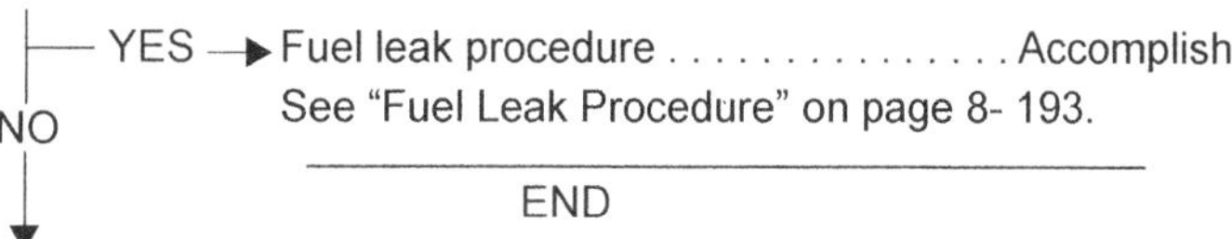

Monitor fuel tank quantities.

L (R) ENG SOV OPEN

Affected FUEL SOV
L or R ENG cb .... Closed

- L engine: (1R8)
- R engine: (1R7)

L (R) ENG SOV CLSD

Engine instruments ....Confirm engine shutdown

Is engine failure confirmed?

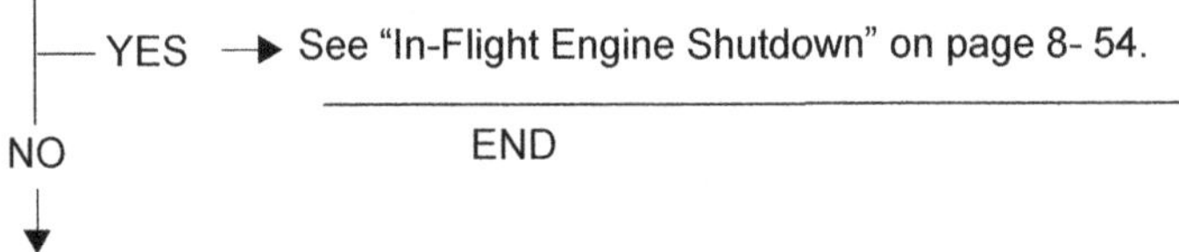

No further action required.

L (R) ENG SOV FAIL

Normal engine operation ....Confirm

Engine instruments .... Monitor

Can engine parameters be maintained within normal limits?

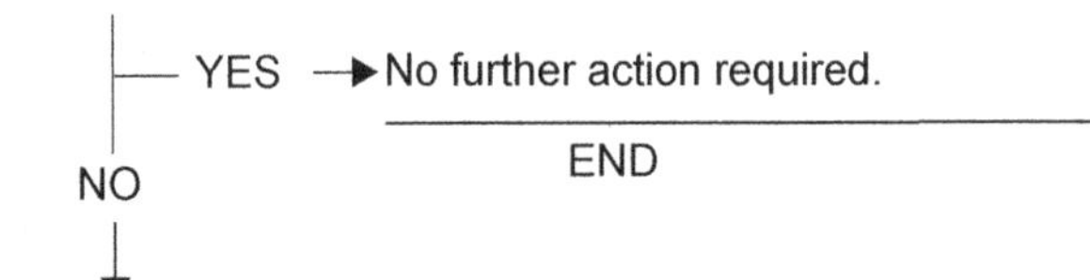

See "In-Flight Engine Shutdown" on page 8- 54.

L (R) FUEL LO PRESS

Affected engine instruments .... Monitor

Is the message accompanied by same side MAIN EJECTOR caution message and/or SCAV EJECTOR caution message?

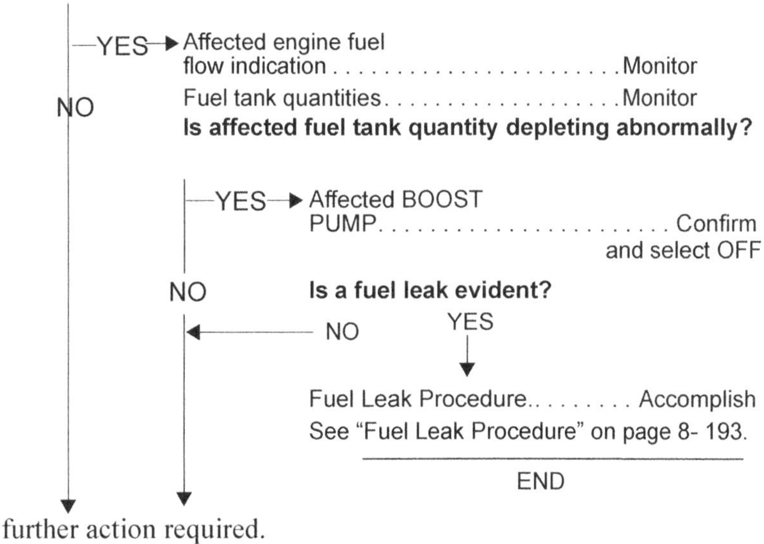

No further action required.

L (R) SCAV EJECTOR

NOTE:
If L/R MAIN EJECTOR messages also show, see "L (R) MAIN EJECTOR and L (R) SCAV EJECTOR" on page 8-182.

Fuel quantity and balance .... Monitor

NOTE:
Collector tank quantity may decrease at high thrust settings. This will be indicated by the fuel quantity indications being displayed amber. If this occurs, reduce pitch attitude and/or thrust on the affected engine to avoid engine flameout.

L (R) MAIN EJECTOR

NOTE:
If L(R) SCAV EJECTOR messages also show, see "L (R) MAIN EJECTOR and L (R) SCAV EJECTOR" on page 8-182.

Left and right boost pumps..................Confirm operating

Affected engine instruments .... Monitor

Fuel tank quantity .... Monitor

and balance if required

Is the center tank quantity increasing or is a leak into the center tank suspected?

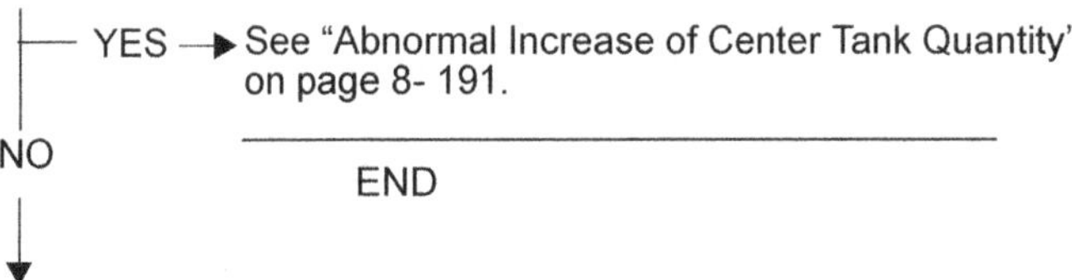

No further action required.

L (R) FUEL PUMP

Affected BOOST PUMP switch.......... Confirm and reset

(select out then back in)

Does the L or R FUEL PUMP caution message persist?

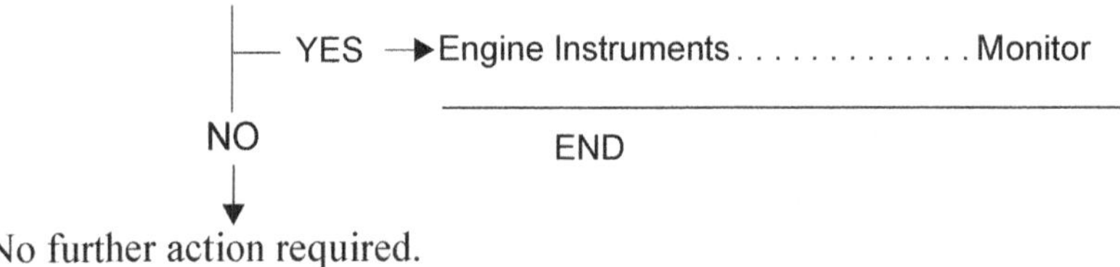

No further action required.

L (R) XFER SOV

Left, right and center fuel tank quantities...........Compare

Left and right fuel tank quantities......Balance, if required

Center tank quantity.. Monitor

Have the L and R XFER SOVs been inhibited, and the center tank quantity increases by more than 150 lbs. or the center tank quantity is more than 600 lbs.?

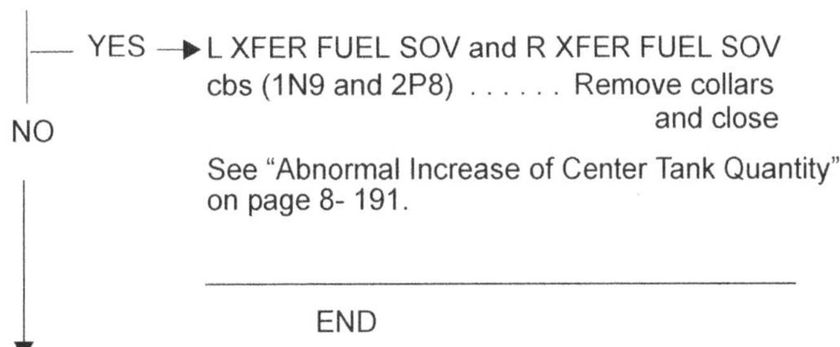

Is the center tank quantity depleting normally with fuel burn?

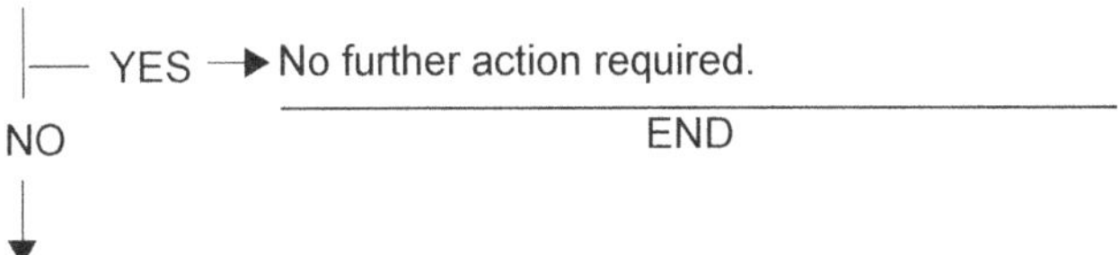

Land immediately at the nearest suitable airport.

XFLOW PUMP

Fuel quantities .... Monitor

Does a fuel imbalance condition exist?

YES → See "Gravity Crossfeed Procedure" on page 8-190.

END

NO ↓

No further action required.

(On the ground, if msg remains, consider fault reset attempt: see page 4-26..

BULK FUEL TEMP

Descend or deviate to warmer air mass and/or increase airspeed.

Engine Instruments .... Monitor

L (R) FUEL LO TEMP

Affected engine instruments .... Monitor

L (R) FUEL FILTER

Affected engine instruments .... Monitor

NOTE:
Fuel filter may be in bypass mode.

Boost Pump Cycling

Center tank quantity readouts .... Monitor

NOTE:
Boost pump cycling not accompanied by a center tank quantity increase above 500 lbs., is acceptable.

Is the center tank quantity increasing abnormally (more than 500 lbs.)?

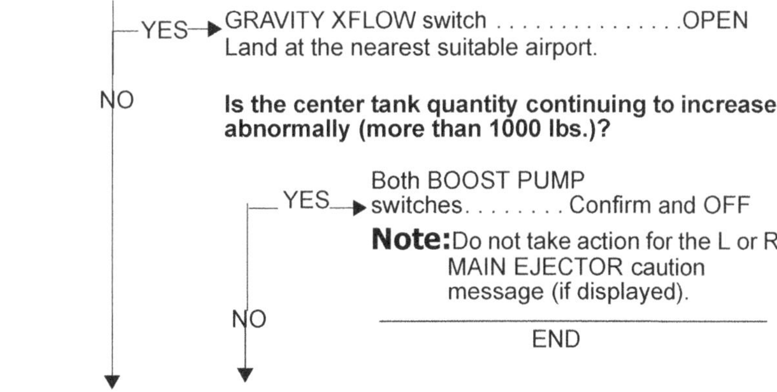

No further action required.

Gravity Crossfeed Procedure

L and R XFLOW switches................................ Select off

XFLOW, AUTO OVERRIDE switch .... Select off

GRAVITY XFLOW switch................................... OPEN

Steady heading sideslip maneuver................ Accomplish

NOTE:
Establish bank angle of 10 degrees down on the low quantity side. Use rudder pedal/trim to maintain a constant heading/course.

NOTE:
During the maneuver, fuel will transfer at a rate of up to 100 lbs. per minute. Fuel tank quantity gauging will be degraded. Accurate fuel indications will occur after 30 seconds of stabilized and coordinated flight.

Fuel tank quantities.. Monitor

Are main tank quantities balanced?

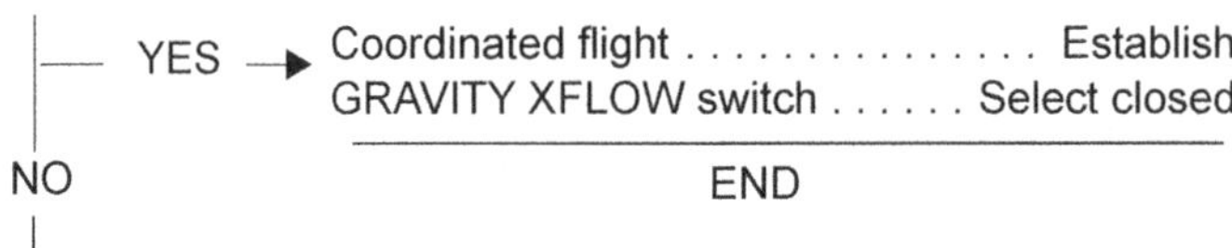

Coordinated flight . . . . Establish
GRAVITY XFLOW switch . . . . . . Select closed

END

Does the fuel imbalance persist and cannot be controlled within limits?

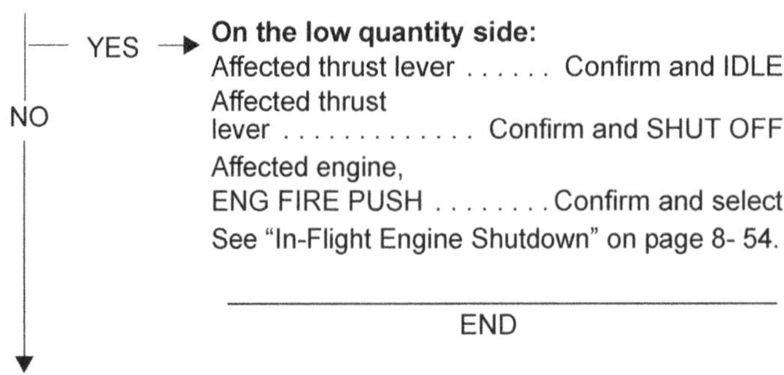

On the low quantity side:
Affected thrust lever . . . . . . Confirm and IDLE
Affected thrust lever . . . . Confirm and SHUT OFF
Affected engine, ENG FIRE PUSH . . . . Confirm and select
See “In-Flight Engine Shutdown” on page 8- 54.

END

No further action required.

Abnormal Increase of Center Tank Quantity

Suspected Leak Into Center Tank

NOTE:
It is acceptable to have a temporary limited fuel variance of up to 300 lb. observed in the center tank during airplane pitch, roll and acceleration maneuvers.

Fuel tank quantity readouts................................. Monitor

On the low quantity side:

Affected thrust lever

(on low quantity side) .... Confirm and IDLE

Land immediately at the nearest suitable airport.

Have the L and R XFER SOVs been inhibited?

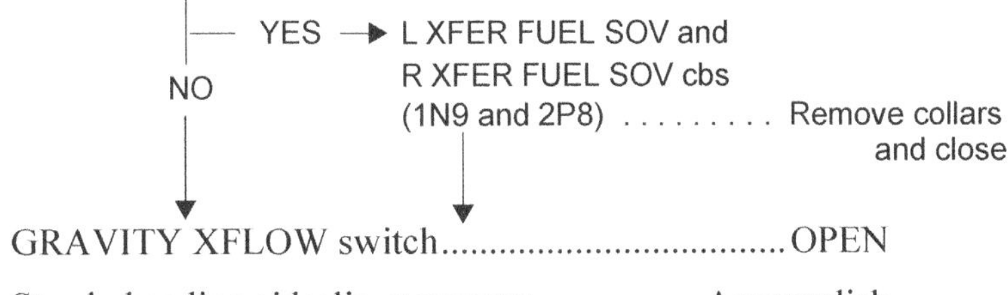

GRAVITY XFLOW switch................................... OPEN

Steady heading sideslip maneuver................ Accomplish

NOTE:
Establish bank angle of 10 degrees down on the low quantity side. Use rudder pedal/trim to maintain a constant heading/course.

NOTE:
During the maneuver, fuel will transfer at a rate of up to 100 lbs. per minute. Fuel tank quantity gauging will be degraded. Accurate fuel indications will occur after 30 seconds of stabilized and coordinated flight.

Fuel tank quantities.. Monitor

CONTINUED ON NEXT PAGE

Are main tank quantities balanced?

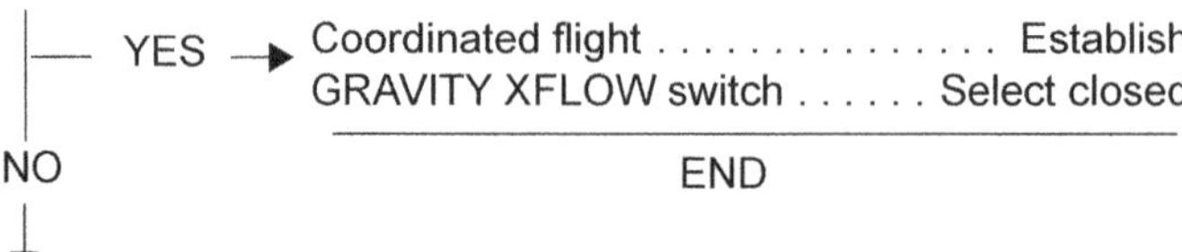

Does the center tank quantity continue to increase (more than 1000 lbs.) or is it abnormally full?

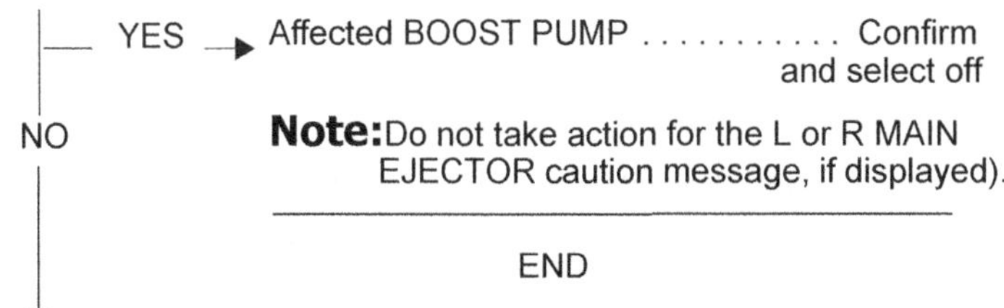

Does the low quantity (affected side) wing tank deplete to 2000 lbs. and center tank quantity is increasing or abnormally full?

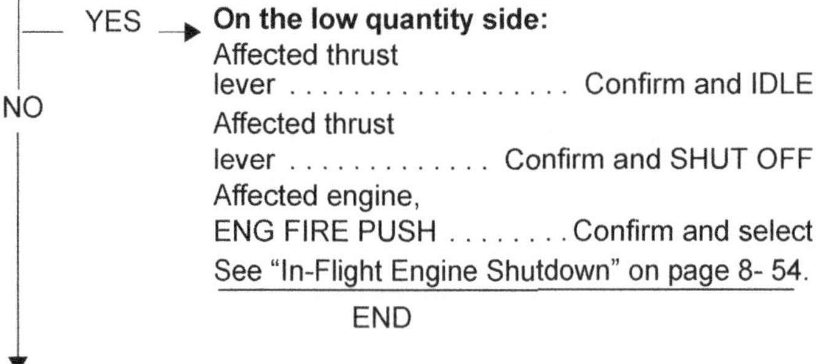

No further action required.

Fuel Leak Procedure

Diversion may be required.

NOTE:
If visibility permits, a visual check from the cabin may enable identification of the leak source.

If a leak is confirmed or suspected:

Land immediately at the nearest suitable airport.

NOTE:
Do not delay landing while attempting to determine location of the leak. Expedite landing if LO FUEL message displayed.

NOTE:
The minimum fuel quantity for go-around is 600 lbs. per wing (with the airplane level) and assuming a maximum airplane climb of 10 degrees nose up.

NOTE:
Do not take action for FUEL IMBALANCE, MAIN EJECTOR, SCAV EJECTOR, or FUEL LO PRESS messages.

Autopilot operation... Monitor

NOTE:
Anticipate an out-of-trim situation when disconnecting the autopilot.

CONTINUED ON NEXT PAGE

Is a leak from a wing tank suspected (wing tank isolation is required)?

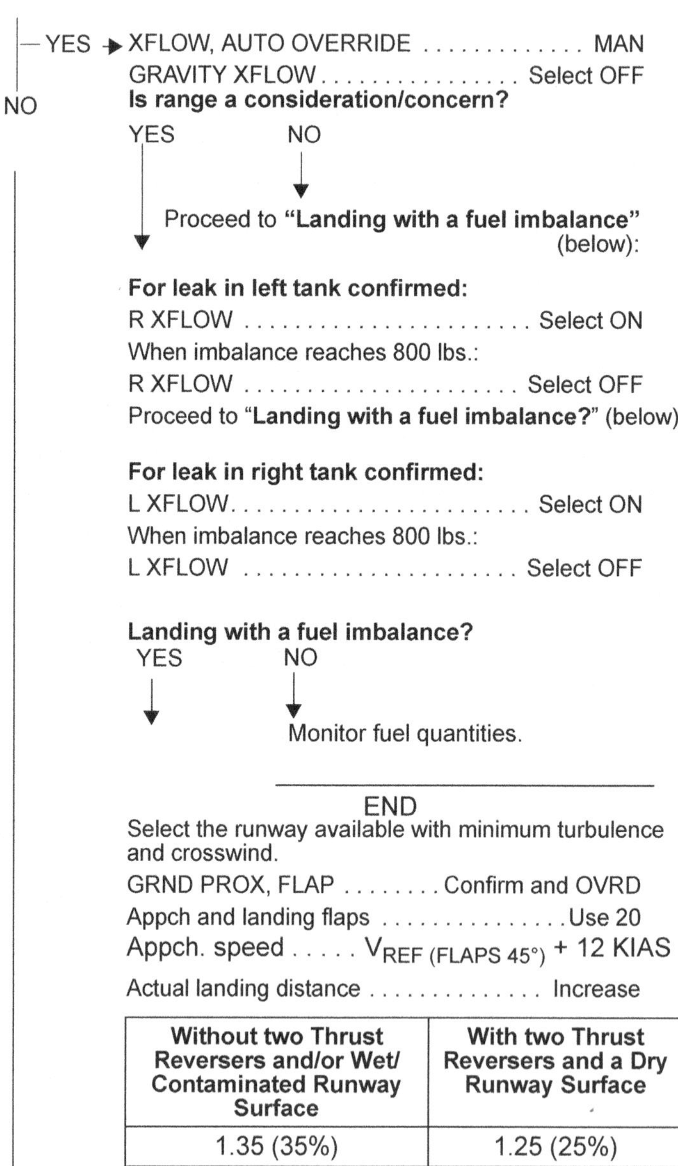

YES → XFLOW, AUTO OVERRIDE .... MAN

GRAVITY XFLOW .... Select OFF

Is range a consideration/concern?

YES NO

NO: Proceed to **"Landing with a fuel imbalance"** (below):

For leak in left tank confirmed:

R XFLOW .... Select ON

When imbalance reaches 800 lbs.:

R XFLOW .... Select OFF

Proceed to "**Landing with a fuel imbalance?**" (below):

For leak in right tank confirmed:

L XFLOW .... Select ON

When imbalance reaches 800 lbs.:

L XFLOW .... Select OFF

Landing with a fuel imbalance?

YES NO

NO: Monitor fuel quantities.

END

Select the runway available with minimum turbulence and crosswind.

GRND PROX, FLAP .... Confirm and OVRD

Appch and landing flaps .... Use 20

Appch. speed ..... $V_{REF\ (FLAPS\ 45°)}$ + 12 KIAS

Actual landing distance .... Increase

Without two Thrust Reversers and/or Wet/ Contaminated Runway Surface	With two Thrust Reversers and a Dry Runway Surface
1.35 (35%)	1.25 (25%)

NO

END

CONTINUED ON NEXT PAGE

Is a leak into the center tank suspected?

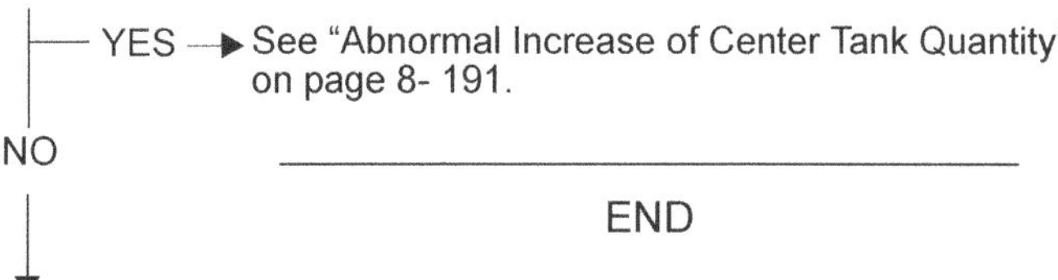

Is a leak from an engine suspected?

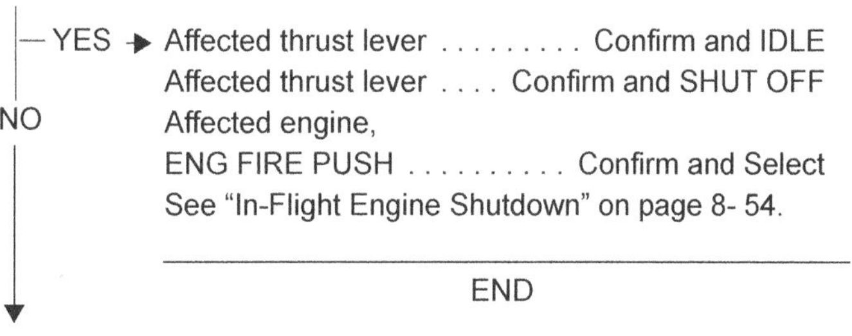

Monitor fuel quantities.

Intentionally Left Blank

Flight Controls Index

Intentionally Left Blank

Flight Controls

Aileron System Jammed

Autopilot	Disengage
Aileron Controls (both)	Release pressure
ROLL DISC	Confirm then PULL & TURN to lock
Airplane control	Transfer to pilot with operative aileron

NOTE:
Roll controllability is reduced. The roll disconnect will result in half feel during airplane handling.

PLT ROLL or CPLT ROLL .... Select operative side

NOTE:
If the PLT ROLL or CPLT ROLL is not selected within 20 seconds of pulling the ROLL DISC, the SPOILERONS ROLL caution message will come on.

Land at the nearest suitable airport.

NOTE:
Select the longest runway available with minimum turbulence and crosswind.

Prior to landing:

GRND PROX, FLAP........................Confirm and OVRD

Landing Flaps ....Use 20°

Approach Speed... Not less than $V_{REF(Flaps\ 45)}$+12 KIAS

CONTINUED ON NEXT PAGE

Actual Landing Distance .... Increase

Without two Thrust Reversers and/or Wet/ Contaminated Runway Surface	With two Thrust Reversers and a Dry Runway Surface
1.35 (35%)	1.25 (25%)

Elevator System Jammed

Autopilot	Disengage
Elevator Controls (both)	Release differential pressure
PITCH DISC handle	Confirm then PULL & TURN to lock
Airplane control	Transfer to pilot with operative elevator

Airspeed .... Not more than 250 KIAS

NOTE:
Pitch controllability is reduced. The pitch disconnect will result in half feel during airplane handling.

NOTE:
Stick pusher is inoperative, if the right side is jammed.

NOTE:
Avoid excessive elevator inputs.

NOTE:
ELEVATOR SPLIT caution message may be displayed if elevator use is aggressive.

Land at the nearest suitable airport.

NOTE:
Select the longest runway available with minimum turbulence and crosswind.

Prior to landing:

GRND PROX, FLAP .... Confirm and OVRD

Landing Flaps .... Use 20°

Approach Speed .... Not less than $V_{REF(Flaps\ 45)}$+12 KIAS

CONTINUED ON NEXT PAGE

Actual Landing Distance .... Increase

Without two Thrust Reversers and/or Wet/ Contaminated Runway Surface	With two Thrust Reversers and a Dry Runway Surface
1.35 (35%)	1.25 (25%)

NOTE:
A slight pitch-up tendency may occur upon selection of reverse thrust. This can be readily corrected by the application of nose-down elevator and/or brakes.

After landing:

If ELEVATOR SPLIT caution message was displayed, airplane structure may be compromised and must be inspected for damage.

Rudder System Jammed

YAW DAMPER DISC .... Select

Rudder pedals .... Overpower

NOTE:
Rudder may be available at higher pedal forces.

NOTE:
Limited travel may be available through the use of rudder trim.

Land at nearest suitable airport. Select the longest runway with minimal turbulence and crosswind.

NOTE:
If the rudder jammed out of neutral position, use aileron and differential thrust to maintain straight flight until touchdown.

Prior to landing:

GRND PROX, FLAP .... Confirm and OVRD

Landing Flaps .... Use 20°

Approach Speed.... Not less than $V_{REF(Flaps\ 45)}$+12 KIAS

Actual Landing Distance .... Increase

Without two Thrust Reversers and/or Wet/ Contaminated Runway Surface	With two Thrust Reversers and a Dry Runway Surface
1.35 (35%)	1.25 (25%)

NOTE:
After touchdown, use differential braking to maintain directional control, as the airplane will turn in the direction of the jammed rudder.

NOTE:
Before using the nose wheel steering tiller ensure that the nose strut is compressed to prevent nose wheel steering failure.

Stabilizer Trim Runaway

Control Wheel . . . . Assume manual control and override runaway

STAB TRIM DISC . . . . Select

Airspeed.......................Not more than 250 KIAS (.70 M)

NOTE:
Autopilot, stabilizer trim and Mach trim are not available.

SEAT BELTS ....ON

Land at nearest suitable airport.

Prior to landing:

GRND PROX, FLAP........................Confirm and OVRD

Landing Flaps ....Use 20°

Approach Speed.... Not less than $V_{REF(Flaps\ 45)}$+12 KIAS

Actual Landing Distance .... Increase

Without two Thrust Reversers and/or Wet/ Contaminated Runway Surface	With two Thrust Reversers and a Dry Runway Surface
1.35 (35%)	1.25 (25%)

NOTE:
A slight pitch-up tendency may occur upon selection of reverse thrust. This can be readily corrected by the application of nose-down elevator and/or brakes.

STALL FAIL

STALL PTCT, PUSHER (left or right) .... OFF

Approach speed .... $V_{REF\ (Flaps\ 45)}$ +10 KIAS minimum

Actual landing distance .... Increase

Without two Thrust Reversers and/or Wet/ Contaminated Runway Surface	With two Thrust Reversers and a Dry Runway Surface
1.15 (15%)	1.10 (10%)

CAUTION:
The low speed awareness cues may represent preset / default settings and should not be relied upon for proximity to stall shaker. increase all reference speeds by 10 kias.

If the Left channel of stall protection system is failed:

Windshear guidance is operative on the copilot's side PFD only.

If the Right channel of stall protection system is failed:

Windshear guidance is operative on the pilot's side PFD only.

If Both channels of stall protection system are failed:

Windshear detection and guidance is inoperative. WINDHSEAR FAIL status message comes on.

PITCH FEEL

Avoid excessive pitch inputs. Use stabilizer trim to alleviate forces on controls.

Prior to landing:

GRND PROX FLAP.........................Confirm and OVRD

Landing Flaps ....Use 20°

Approach speed.........$V_{REF\ (Flaps\ 45)}$ +12 KIAS minimum

Actual landing distance... Increase

Without two Thrust Reversers and/or Wet/ Contaminated Runway Surface	With two Thrust Reversers and a Dry Runway Surface
1.35 (35%)	1.25 (25%)

RUD LIMITER

Is Rudder limiter failed at or near full travel?

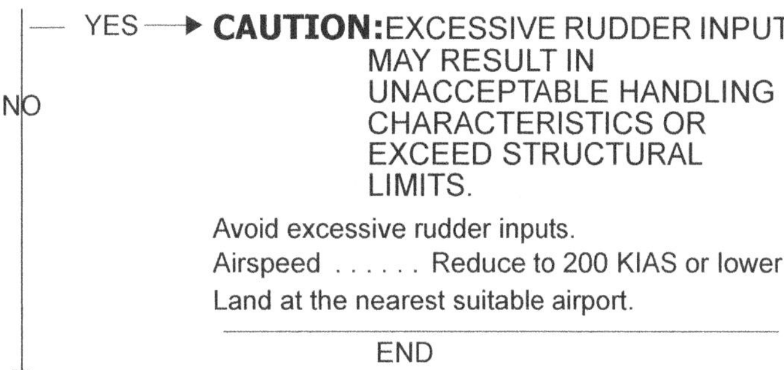

Rudder limiter failed at an intermediate or most limited position:

CAUTION:
Rudder travel is limited. additional aileron input may be required to maintain directional control.

Land at the nearest suitable airport.

Select the runway available with minimum turbulence and crosswind.

Prior to landing:

GRND PROX FLAP........................Confirm and OVRD

Landing Flaps ....Use 20°

Approach speed .... $V_{REF\ (Flaps\ 45)}$ +12 KIAS Minimum

Actual landing distance......................................Increase

Without two Thrust Reversers and/or Wet/ Contaminated Runway Surface	With two Thrust Reversers and a Dry Runway Surface
1.35 (35%)	1.25 (25%)

ELEVATOR SPLIT

Autopilot .... Disconnect

Airspeed .... Reduce to 200 KIAS or lower

Avoid Excessive elevator input.

NOTE:
Use pitch trim as necessary to minimize elevator deflections. Use aileron control as necessary to stabilize the airplane.

Land at the nearest suitable airport.

Prior to landing:

NOTE:
Controllability is reduced. Select the longest runway available with minimum turbulence and crosswind.

GRND PROX FLAP.........................Confirm and OVRD

Landing Flaps ....Use 20°

Approach speed........ $V_{REF\ (Flaps\ 45)}$ +12 KIAS Minimum

Actual landing distance.. Increase

Without two Thrust Reversers and/or Wet/ Contaminated Runway Surface	With two Thrust Reversers and a Dry Runway Surface
1.35 (35%)	1.25 (25%)

After landing:

Airplane structure may be compromised and must be inspected for damage.

Aileron PCU Runaway

NOTE:
The PLT ROLL or CPLT ROLL light on the glareshield indicates the operative aileron.

NOTE:
If the PLT ROLL or CPLT ROLL light is not on, identify the failed side by verifying aileron response to handwheel movements in the FLIGHT CONTROLS synoptic page.

Aileron control.. Transfer to pilot with operative aileron

Autopilot.. Disengage

ROLL DISC.............................. Confirm then PULL and TURN to lock

NOTE:
Following disconnect, the inoperative side handwheel will go hardover in the direction of the failure.

NOTE:
Roll controllability is reduced. The roll disconnect will result in half feel during airplane handling.

PLT ROLL or CPLT ROLL ....Select operative side

NOTE:
If the PLT ROLL or CPLT ROLL switch is not selected within 20 seconds of pulling the ROLL DISC handle, the SPOILERONS ROLL caution message will come on.

Land at the nearest suitable airport.

CONTINUED ON NEXT PAGE

Prior to landing:

NOTE:
Select the longest runway available with minimum turbulence and crosswind.

GRND PROX FLAP.........................Confirm and OVRD

Landing Flaps ....Use 20°

Approach speed.........$V_{REF\ (Flaps\ 45)}$ +12 KIAS minimum

Actual landing distance.......................................Increase

Without two Thrust Reversers and/or Wet/ Contaminated Runway Surface	With two Thrust Reversers and a Dry Runway Surface
1.35 (35%)	1.25 (25%)

MACH TRIM

STAB TRIM, CH1 and CH2 .... Engage

NOTE:
With a Mach trim failure, do not exceed 250 KIAS (0.70M) unless the autopilot is engaged and operating normally.

NOTE:
At least one STAB TRIM channel engagement is required for MACH TRIM operation.

MACH TRIM .... Engage

STAB TRIM

STAB TRIM CH 1 and CH2 .... Engage

Does the STAB TRIM caution message disappear?

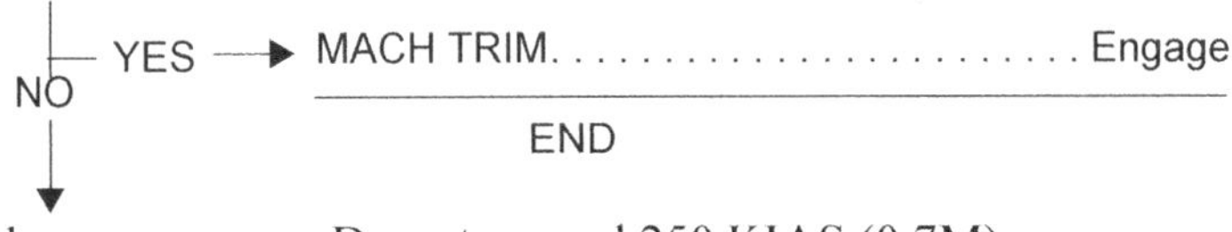

Airspeed........................Do not exceed 250 KIAS (0.7M)

NOTE:
Autopilot, stabilizer trim and mach trim are not available.

NOTE:
Delay changing airspeed or configuration as long as possible to minimize the out-of-trim condition.

NOTE:
For nose down trim condition when significant pull forces are required, selecting FLAPS to 1 or 8 will reduce the control forces.

NOTE:
Category II operation may be affected by this failure. Review the equipment requirements found in chapter 3 of this manual.

Prior to landing:

GRND PROX FLAP.........................Confirm and OVRD

Landing Flaps ....Use 20°

Approach speed ....$V_{REF\ (Flaps\ 45)}$ +12 KIAS minimum

Actual landing distance.. Increase

Without two Thrust Reversers and/or Wet/ Contaminated Runway Surface	With two Thrust Reversers and a Dry Runway Surface
1.35 (35%)	1.25 (25%)

Aileron or Rudder Trim Runaway

Aileron (or rudder).. Overpower

Affected trim.. Neutral

Does the trim runaway persist?

YES → Applicable trim cb . . . . Open

Rudder trim cb (2F2)

Aileron trim cb (2F3)

Note: Use of autopilot may alleviate aileron trim problem.

END

NO ↓

Operate the trim with caution for the remainder of the flight

STAB TRIM LIMIT

NOTE:
An out of trim condition may exist when disconnecting the autopilot.

Autopilot.. Disconnect

NOTE:
A slight back force in pitch may be required.

Airplane .... Retrim

Leave icing conditions.

NOTE:
Icing conditions exist in-flight at a TAT of 10°C (50°F) or below, and visible moisture in any form is encountered (such as clouds, rain, snow, sleet or ice crystals), except when the SAT is -40°C (-40°F) or below.

SPOILERONS ROLL

On the side with the operative control:

PLT ROLL or CPLT ROLL .... Select to operative side

FLT SPLR DEPLOY

FLIGHT SPOILER lever .... RETRACT

FLIGHT SPOILERS Lever Jam (Spoilers Deployed)

Thrust .... Set as required

Prior to landing:

Approach speed ....$V_{REF\ (Flaps\ 45)}$ +10 KIAS minimum

NOTE:
The landing distance factors below are based upon loss of all multi-functional spoilers.

Actual landing distance....................................... Increase

Without two Thrust Reversers and/or Wet/ Contaminated Runway Surface	With two Thrust Reversers and a Dry Runway Surface
1.35 (35%)	1.25 (25%)

FLAPS FAIL

NOTE:
If SLATS FAIL message also shows, see "FLAPS FAIL and SLATS FAIL" on page 8-220.

SLATS/FLAPS lever .... Select last position and then re-select

Has the FLAPS FAIL condition been fixed?

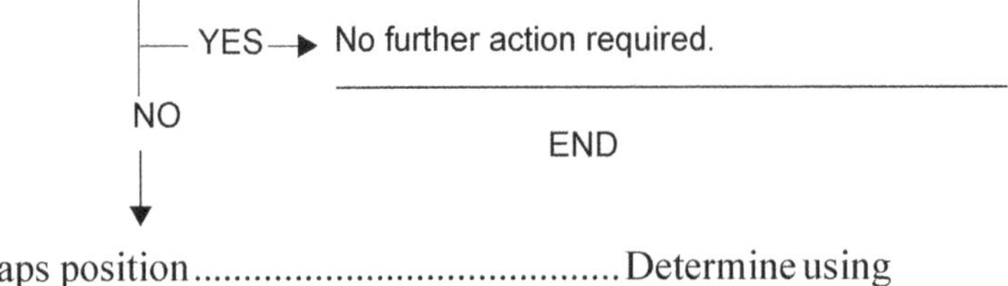

Flaps position .... Determine using F/CTL synoptic page

Has the flaps failure occurred at a detented position?

YES → SLATS/FLAPS lever .Select detented position. Do not attempt to operate flaps until further advised.

Maximum enroute airspeed. . . . . V_{FE} for detented flap position

Altitude . . . . .Not more than 15,000 feet

Note: If the flaps are confirmed retracted (0), reduction of cruise airspeed/altitude is not required.

Proceed to **"Prior to landing"** (see below).

END

NO ↓

SLATS/FLAPS lever .... Select closest detented position less than actual flap position. Do not attempt to operate flaps any further.

Maximum airspeed .... V_{FE} for next greater flap setting from failed position

CONTINUED ON NEXT PAGE

Altitude .... Not more than 15,000 feet

Prior to landing:

GRND PROX FLAP........................Confirm and OVRD

HYDRAULIC 1, 2 and 3B pump ....ON

Are flaps confirmed retracted (at 0°)?

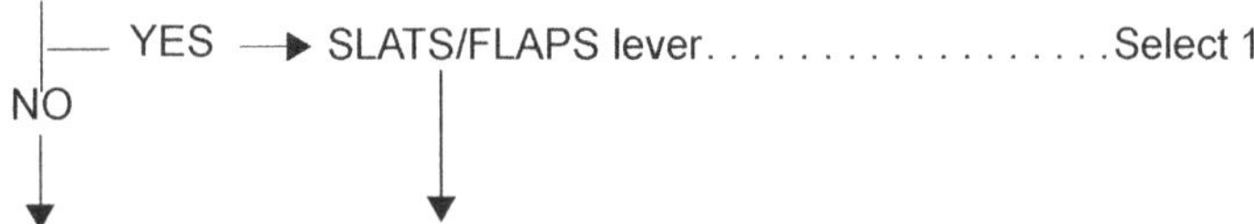

- Final Approach speed$V_{REF\ (FLAPS\ 45°)} + \Delta V_{REF}$ from the following table.

ΔV_{REF}		
Flaps Position	**Slats Position**	
	20-24	25
0-7	24	N/A
8-19	18	N/A
20-29	12	N/A
30-44	N/A	8
45	N/A	0

NOTE:
Windshear escape guidance is inoperative.

NOTE:
For flaps failed at less than full extension, a higher than normal pitch attitude will be required during the approach. The visual cues may be misleading. Use any available glidepath reference (ILS, FMS VNAV, VASI, etc.) to maintain the proper glidepath control.

CONTINUED ON NEXT PAGE

Actual landing distance....................................... Increase:

Final Approach Speed ΔV_{REF} (KTS)	**Actual Landing Distance Factor (Without two Thrust Reversers and/or Wet/Contaminated Runway Surface)**	**Actual Landing Distance Factor (With two Thrust Reversers and a Dry Runway Surface)**
24	1.45	1.35
18	1.40	1.30
12	1.35	1.25
8	1.30	1.20

CONTINUED ON NEXT PAGE

Is ΔV_{REF} 18 or 24 knots?

YES ▶ Max landing weight . . . . Determine using the following table and correct for wind.

OAT		Airport Pressure Altitude (Feet)					
		0	**2000**	**4000**	**6000**	**8000**	**10000**
°C	**°F**	**Landing Weight (lbs.) Associated with max. tire speed**					
-40	-40	85,980	85,980	85,980	85,980	85,980	85,980
-20	-4	85,980	85,980	85,980	85,980	85,980	85,980
0	32	85,980	85,980	85,980	85,980	85,980	80,811
20	68	85,980	85,980	85,980	85,980	81,939	74,907
40	104	85,980	85,980	85,980	83,393	76,127	69,668

Wind corrections:

Increase landing weight by 4410 lbs per 10 kts headwind.

Decrease landing weight by 14,330 lbs per 10 kts tailwind

Notes:
1. The actual landing weight must not exceed the corrected maximum landing weight due to tire speed.
2. A slight pitch-up tendency may occur upon selection of reverse thrust. This can be readily corrected by the application of nose-down elevator and/or brakes.

END

NO ▼

No further action required

SLATS FAIL

NOTE:
If FLAPS FAIL message also shows, see "FLAPS FAIL and SLATS FAIL" on page 8-220.

SLATS/FLAPS lever .... Select last position and then re-select.

Has the SLATS FAIL condition been fixed?

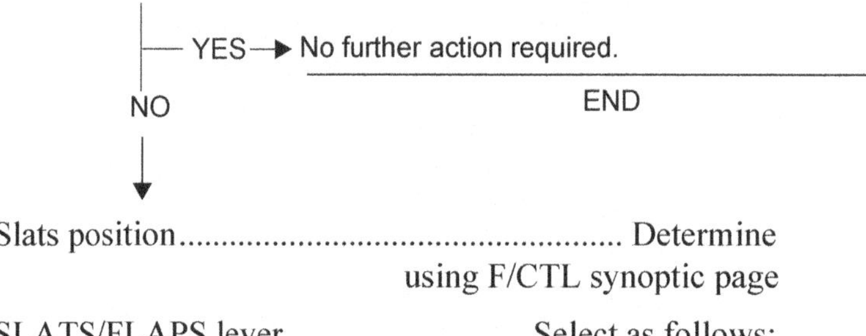

Slats position .... Determine using F/CTL synoptic page

SLATS/FLAPS lever .... Select as follows:

Do not attempt to operate flaps until further advised.

Slats Position	Select SLATS/FLAPS Lever to Position
0-19	0
20-24	1
25	20

Max enroute airspeed .... VFE for detented flap position

NOTE:
With flaps 0, do not exceed 230 KIAS unless slats are confirmed retracted (0).

Altitude .... Not more than 15,000 feet

CONTINUED ON NEXT PAGE

NOTE:
If the flaps and the slats are confirmed retracted (0), reduction of cruise airspeed/altitude is not required.

Prior to landing:

GRND PROX, FLAP ....Confirm and OVRD

HYDRAULIC 1, 2 and 3B pump ....ON

SLATS/FLAPS lever ....Select 45

Final Approach speed ....$V_{REF\ (FLAPS\ 45°)} + \Delta V_{REF}$ from the following table.

ΔV_{REF}		
Flaps Position	**Slats Position**	
	0-24	25
45	10	0

NOTE:
Windshear escape guidance is inoperative.

NOTE:
For slats failed at less than full extension, a lower than normal pitch attitude will be required during the approach. The visual curs may be misleading. Use any available glidepath reference (ILS, FMS, VNAV, VASI, etc.) to maintain the proper glidepath control.

Actual landing distance...

Final Approach Speed ΔV_{REF} (KTS)	**Actual Landing Distance Factor (Without two Thrust Reversers and/or Wet/Contaminated Runway Surface)**	**Actual Landing Distance Factor (With two Thrust Reversers and a Dry Runway Surface)**
10	1.30	1.25

FLAPS FAIL and SLATS FAIL

SLATS/FLAPS lever .... Select last position and then re-select

Has the FLAPS FAIL and SLATS FAIL condition been fixed?

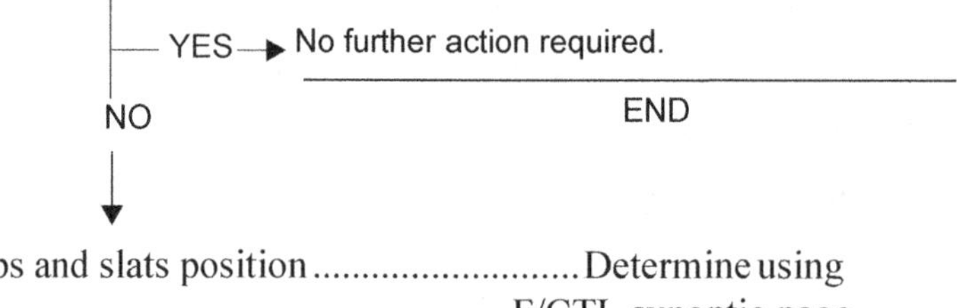

Flaps and slats position .... Determine using F/CTL synoptic page

Has the flaps failure occurred at a detented position?

YES → SLATS/FLAPS lever. Select detented position. Do not attempt to operate flaps until further advised.

Maximum enroute airspeed . . . . V_{FE} for detented flap position

Note: With flaps 0, do not exceed 230 KIAS unless slats are confirmed retracted (0).

Altitude. . . . . Not more than 15,000 feet

Note: If the flaps and slats are confirmed retracted (0), reduction of cruise airspeed/altitude is not required.

Proceed to **"Prior to landing"** (see below).

END

NO ↓

SLATS/FLAPS lever .... Select closest detented position less than actual flap position. Do not attempt to operate flaps any further.

Maximum airspeed .... V_{FE} for next greater flap setting from failed position.

Altitude .... Not more than 15,000 feet

CONTINUED ON NEXT PAGE

Prior to landing:

GRND PROX FLAP........................Confirm and OVRD

HYDRAULIC 1, 2 and 3B pump ....ON

Final Approach speed ....$V_{REF\ (FLAPS\ 45°)} + \Delta V_{REF}$ from the following table.

ΔV_{REF}			
Flaps Position	**Slats Position**		
	0-19	20-24	25
0-7	40	24	24
8-19	30	18	18
20-29	30	12	12
30-44	24	24	8
45	10	10	0

NOTE:
Windshear escape guidance is inoperative.

NOTE:
For flaps failed at less than full extension, a higher than normal pitch attitude will be required during the approach. The visual cues may be misleading. Use any available glidepath reference (ILS, FMS VNAV, VASI, etc.) to maintain the proper glidepath control.

Actual landing distance....................................... Increase

Final Approach Speed ΔV_{REF} (KTS)	**Actual Landing Distance Factor (Without two Thrust Reversers and/or Wet/Contaminated Runway Surface)**	**Actual Landing Distance Factor (With two Thrust Reversers and a Dry Runway Surface)**
40	1.70	1.60
30	1.55	1.50
24	1.45	1.35
18	1.40	1.30
12	1.35	1.25
10	1.30	1.25
8	1.30	1.20

CONTINUED ON NEXT PAGE

Is ΔV_{REF} 40 knots?

YES ▶ Max landing weight . . . . Determine using the following table and correct for wind.

OAT		Airport Pressure Altitude (Feet)					
		0	**2000**	**4000**	**6000**	**8000**	**10000**
°C	**°F**	**Landing Weight (lbs.) Associated with max. tire speed**					
-40	-40	85,980	85,980	85,980	85,980	85,980	79,069
-20	-4	85,980	85,980	85,980	85,980	79,063	71,807
0	32	85,980	85,980	85,980	79,301	72,040	65,331
20	68	85,980	85,980	79,861	72,630	65,962	59,752
40	104	85,980	80,790	73,546	66,891	60,580	54,748

Wind corrections:

Increase landing weight by 4410 lbs per 10 kts headwind.

Decrease landing weight by 15,430 lbs per 10 kts tailwind

Notes:
1. The actual landing weight must not exceed the corrected maximum landing weight due to tire speed.
2. A slight pitch-up tendency may occur upon selection of reverse thrust. This can be readily corrected by the application of nose-down elevator and/ or brakes.

END

NO ▼

CONTINUED ON NEXT PAGE

Is ΔV_{REF} between 18, 24, or 30 knots?

YES ▶ Max landing weight . . . . Determine using the following table and correct for wind.

OAT		Airport Pressure Altitude (Feet)					
		0	**2000**	**4000**	**6000**	**8000**	**10000**
°C	**°F**	**Landing Weight (lbs.) Associated with max. tire speed**					
-40	-40	85,980	85,980	85,980	85,980	85,980	85,980
-20	-4	85,980	85,980	85,980	85,980	85,980	81,532
0	32	85,980	85,980	85,980	85,980	82,179	74,912
20	68	85,980	85,980	85,980	83,198	75,835	69,168
40	104	85,980	85,980	84,307	76,871	70,101	63,819

Wind corrections:

Increase landing weight by 4,410 lbs per 10 kts headwind.

Decrease landing weight by 14,770 lbs per 10 kts tailwind

Notes:
1. The actual landing weight must not exceed the corrected maximum landing weight due to tire speed.
2. A slight pitch-up tendency may occur upon selection of reverse thrust. This can be readily corrected by the application of nose-down elevator and/or brakes.

END

NO

No further action required.

SLATS/FLAPS Lever Jammed or Disconnected

Is an undesirable slat/flap configuration present?

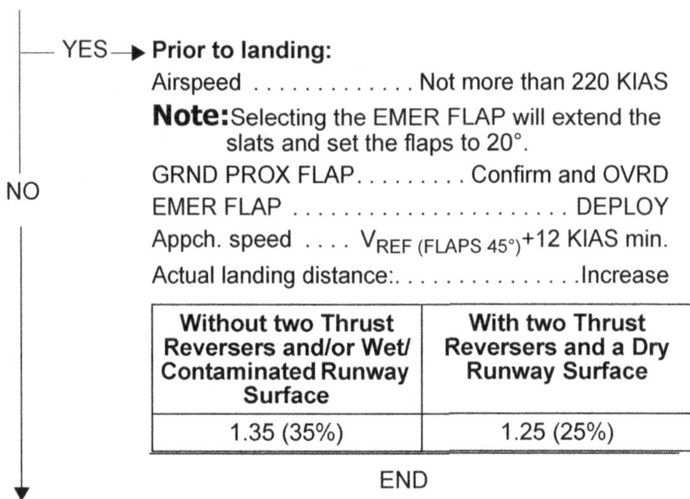

YES → **Prior to landing:**

Airspeed . . . . Not more than 220 KIAS

Note: Selecting the EMER FLAP will extend the slats and set the flaps to 20°.

GRND PROX FLAP. . . . . Confirm and OVRD

EMER FLAP . . . . DEPLOY

Appch. speed . . . . $V_{REF\ (FLAPS\ 45°)}$+12 KIAS min.

Actual landing distance:. . . . . .Increase

Without two Thrust Reversers and/or Wet/ Contaminated Runway Surface	With two Thrust Reversers and a Dry Runway Surface
1.35 (35%)	1.25 (25%)

END

NO ↓

No further action required.

IB FLT SPLRS

Are the flight spoilers deployed?

Airplane altitude .... 26,000 feet maximum

Is the IB SPOILERONS caution message also displayed?

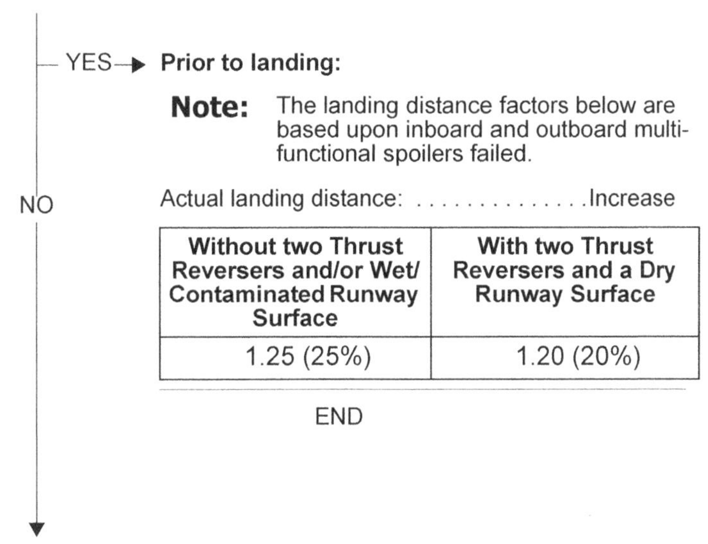

Without two Thrust Reversers and/or Wet/ Contaminated Runway Surface	With two Thrust Reversers and a Dry Runway Surface
1.25 (25%)	1.20 (20%)

No further action required.

OB FLT SPLRS

Are the flight spoilers deployed?

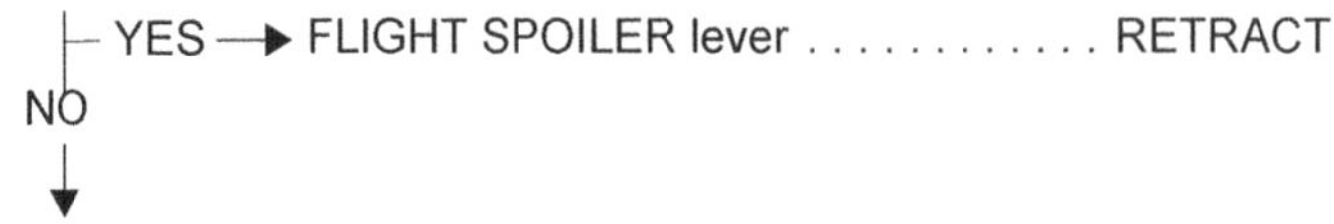

Airplane altitude .... 26,000 feet maximum

Is the OB SPOILERONS caution message also displayed?

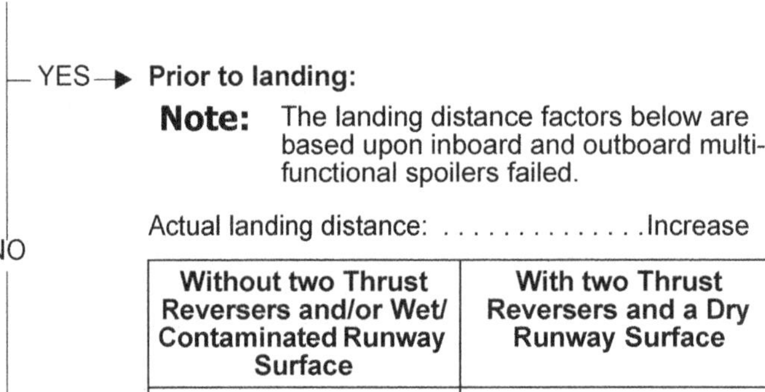

Prior to landing:

Note: The landing distance factors below are based upon inboard and outboard multi-functional spoilers failed.

Actual landing distance: .... Increase

Without two Thrust Reversers and/or Wet/ Contaminated Runway Surface	With two Thrust Reversers and a Dry Runway Surface
1.25 (25%)	1.20 (20%)

END

No further action required.

IB SPOILERON

NOTE:
Select the runway available with minimum crosswind.

NOTE:
The landing distance factors below are based upon inboard and outboard multi-functional spoilers failed.

Is the IB FLT SPLRS caution message also displayed?

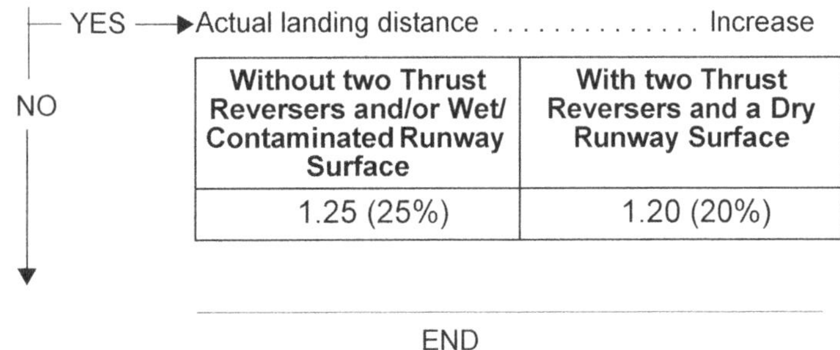

Without two Thrust Reversers and/or Wet/ Contaminated Runway Surface	With two Thrust Reversers and a Dry Runway Surface
1.25 (25%)	1.20 (20%)

No further action required.

(On the ground, if message remains with IB (OB) FLT SPLR FAULT status message, consider fault reset attempt. See page 4-26).

OB SPOILERON

NOTE:
Select the runway available with minimum crosswind.

NOTE:
The landing distance factors below are based upon inboard and outboard multi-functional spoilers failed.

Is the OB FLT SPLRS caution message also displayed?

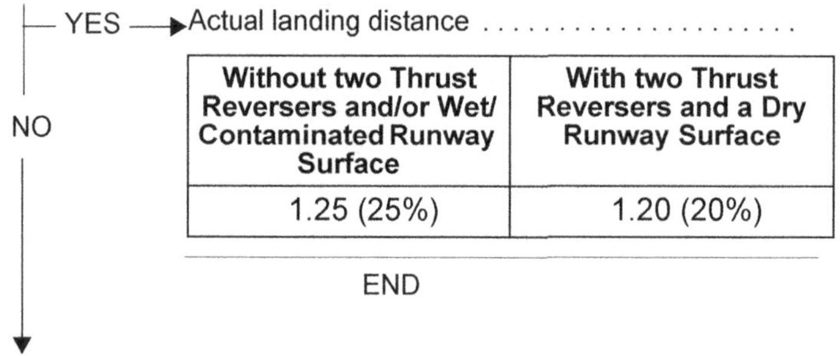

Without two Thrust Reversers and/or Wet/ Contaminated Runway Surface	With two Thrust Reversers and a Dry Runway Surface
1.25 (25%)	1.20 (20%)

No further action required.

(On the ground, if message remains with IB (OB) FLT SPLR FAULT status message, consider fault reset attempt. See page 4-26.)

GND SPLR DEPLOY

GND LIFT DUMPING........................... MAN DISARM

Prior to landing:

GRND PROX, FLAP........................Confirm and OVRD

Landing Flaps ....Use 20°

Approach speed .... $V_{REF\ (FLAPS\ 45°)}$ + 12 KIAS min.

NOTE:
Select the runway with minimum crosswind.

NOTE:
The landing distance factors below are based upon complete loss of ground lift dumping.

Actual landing distance.. Increase

Without two Thrust Reversers and/or Wet/ Contaminated Runway Surface	With two Thrust Reversers and a Dry Runway Surface
1.45 (45%)	1.35 (35%)

After touchdown:

GND LIFT DUMPING.................................. MAN ARM

IB GND SPLRS

GND LIFT DUMPING........................... MAN DISARM

Prior to landing:

Actual landing distance.. Increase

Without two Thrust Reversers and/or Wet/ Contaminated Runway Surface	With two Thrust Reversers and a Dry Runway Surface
1.30 (30%)	1.25 (25%)

After touchdown:

GND LIFT DUMPING.................................. MAN ARM

OB GND SPLRS

GND LIFT DUMPING........................... MAN DISARM

Prior to landing:

Actual landing distance...

Without two Thrust Reversers and/or Wet/ Contaminated Runway Surface	With two Thrust Reversers and a Dry Runway Surface
1.30 (30%)	1.25 (25%)

After touchdown:

GND LIFT DUMPING.................................. MAN ARM

GLD UNSAFE

Choose a scenario

On the ground:

External contributing factors:

- One or both radio altimeter antennas covered with moisture (typical after de-icing or heavy rain).
- Aircraft positioned over a surface with ice, snow or standing water/de-ice fluid.
- Aircraft positioned over surfaces with high metallic content in the under-structure (bridge, tunnel, underground tanks, water drainage system etc.).

The condition is likely temporary. Continue taxi. The message will likely go away as the aircraft is relocated over a different surface.

Does the message persist prior to takeoff?

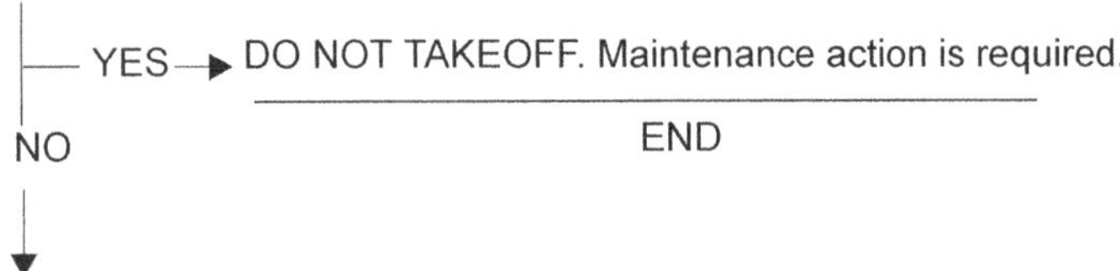

Perform a radio altimeter test on both radio altimeters. If the tests are successful, no further action is required.

CONTINUED ON NEXT PAGE

In flight:

GND LIFT DUMPING........................... MAN DISARM

NOTE:
The landing distance factors below are based upon complete loss of ground lift dumping.

Actual landing distance.. Increase

Without two Thrust Reversers and/or Wet/ Contaminated Runway Surface	With two Thrust Reversers and a Dry Runway Surface
1.35 (35%)	1.25 (25%)

After touchdown:

GND LIFT DUMPING.................................. MAN ARM

GLD NOT ARMED

GND LIFT DUMPING.................................. MAN ARM

Is the GLD NOT ARMED caution message still displayed?

YES → **Prior to landing:**

Note: The landing distance factors below are based upon complete loss of ground lift dumping.

Actual landing distance . . . . Increase

Without two Thrust Reversers and/or Wet/ Contaminated Runway Surface	With two Thrust Reversers and a Dry Runway Surface
1.35 (35%)	1.25 (25%)

After touchdown:

FLIGHT SPOILER lever. . . . . Select MAX deploy

END

NO

No further action required.

Air Conditioning & Pressurization Index

Intentionally Left Blank

Air Conditioning & Pressurization

General

Loss of cabin pressure can be caused by a number of different conditions. Depending on the actual situation, the rate of pressure loss may vary considerably. It is important that the flight crew, after completing the memory items, take the time to carefully assess the actual cabin rate and altitude prior to initiating any subsequent action. In the case of a rapid or explosive decompression, immediate crew action is required.

NOTE:

An explosive decompression can be readily identified. It is usually associated with an explosive noise, flying debris that may affect vision for several seconds, fogging and subsequent rapid drop in cabin temperature.

The time of useful consciousness (effective performance time) is the period of time between the interruption of the oxygen supply or exposure to an oxygen-poor environment and the time when a pilot is unable to perform flying duties effectively, such as donning an oxygen mask and descending to a safe altitude. The following table shows these times and should be used as a guide only as the times are based on healthy individuals at rest in a hyperbaric chamber. After rapid decompression, reverse diffusion (fulminating hypoxia) occurs when oxygen is forced from the lungs due to rapid expansion. The result is acute and immediate hypoxia as can be see by the 50% reduction in time of useful consciousness.

TIMES OF USEFUL CONSCIOUSNESS AT VARIOUS ALTITUDES		
Standard Ascent Rate		**After Rapid Decompression**
Altitude (FT.)	**Time**	**Time**
18,000	20-30 min.	10-15 min.
22,000	10 min.	5 min.
25,000	3-5 min.	1.5-3.5 min.
28,000	2.5-3 min.	1.25-1.5 min.
30,000	1-2 min.	30-60 sec.
35,000	30-60 sec.	15-30 sec.
40,000	15-20 sec.	7-10 sec.
43,000	9-12 sec.	5 sec.

Loss of Cabin Pressure - Memory Items Expanded

Pilot Flying	Pilot Not Flying
Verifies indications and calls: **"Confirmed, memory items"** Dons oxygen mask, selects MASK and establishes communication.	Checks cabin status, cancels warning and calls: **"Loss of cabin pressure."** Dons oxygen mask, selects MASK and establishes communication. Proceeds with the memory items as follows: Checks bleed configuration to ensure that bleeds and PACKS are set properly. If control is not regained: Calls: **"Memory items complete"**
Both pilots: Carefully check cabin rate of climb or descent, cabin altitude and position of discharge valves to determine level of pressurization control. If control **has been** regained, the PF call for this checklist.	
If control **has not been** regained, consider the requirement for an emergency descent. Based on the rate of cabin climb, a more moderate descent profile may be acceptable. If an emergency descent is deemed necessary, clearly state intentions and proceed with the "Emergency Descent" memory items and complete the checklist See "Emergency Descent Procedure" on page 8- 237.	

Cabin Alt (Red)

Emergency Descent Procedure

Cabin Alt (Red)

Oxygen masks	Don, set to 100%
Crew communication	Establish

See emergency checklist card or continue below.

PASS SIGNS switch(es) .... ON

Descent .... Initiate to 10,000 ft. or lowest safe altitude

Thrust levers .... IDLE

FLIGHT SPOILERS lever .... MAX

PASS OXY switch .... Confirm and ON if CABIN ALT is approaching 14,000 ft.

Is structural damage suspected?

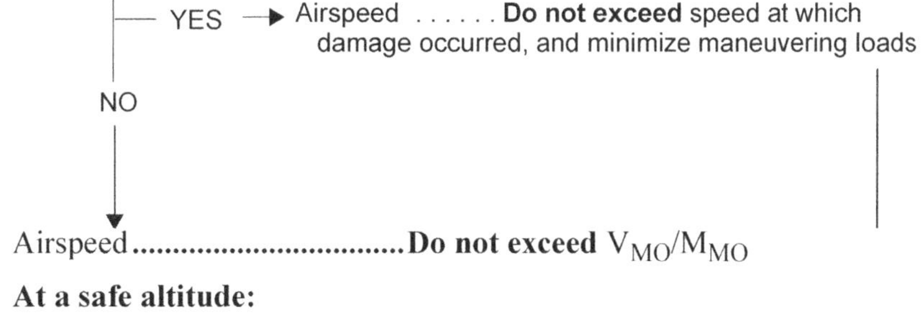

Airspeed .... **Do not exceed** V_{MO}/M_{MO}

At a safe altitude:

Oxygen and masks .... As required

CAUTION:

Closing the doors on the mask stowage compartments or pressing "reset" will stop the flow of oxygen to the masks.

CONTINUED ON NEXT PAGE

NOTE:
If supplemental crew oxygen is still required, setting masks to normal (N) will reduce consumption.

NOTE:
If the PASS OXY system is in use, cabin temperatures will increase and there may be an associated hot/burning smell from the chemical process in the oxygen generators.

Are any DOOR message(s) or hatch unsafe conditions present?

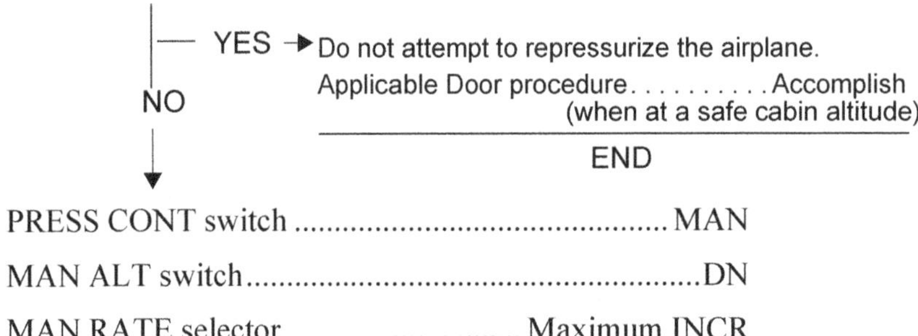

PRESS CONT switch .... MAN

MAN ALT switch...DN

MAN RATE selector .... Maximum INCR

Has control of cabin pressurization been regained?

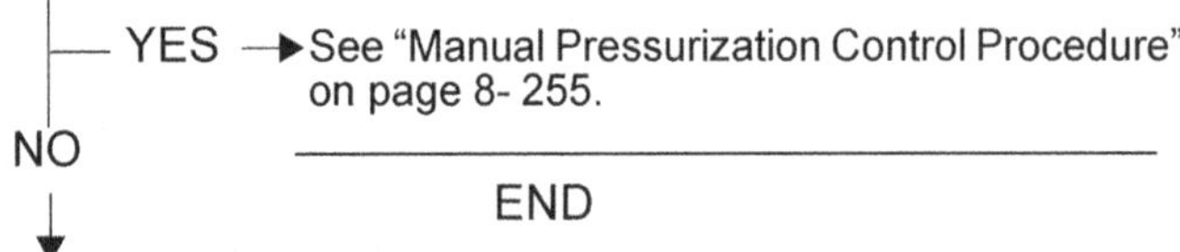

See "Unpressurized Flight Procedure" on page 8- 256.

DIFF PRESS

NOTE:
For altitudes above 30,000 feet, if climbing at a rate greater than 1,500 FPM, a DIFF PRESS warning message may come on.

NOTE:
DIFF PRESS warning message can be posted even if the pressure values on the ECS synoptic page are green.

PRESS CONT switchlight .... MAN

MAN ALT .... UP

MAN RATE... INCR as required

Has control of cabin pressurization been regained?

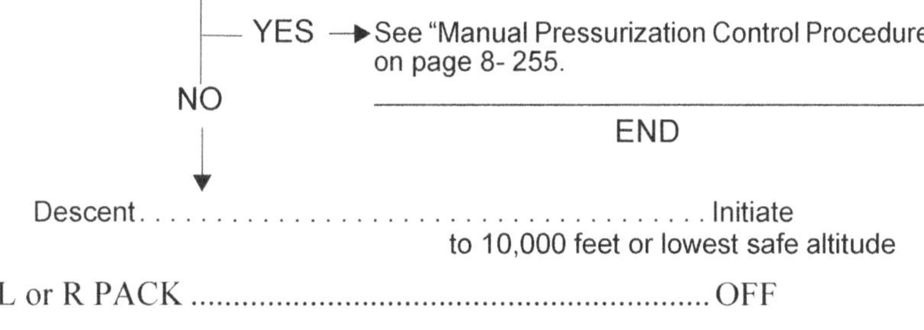

L or R PACK .... OFF

Does the DIFF PRESS warning message persist?

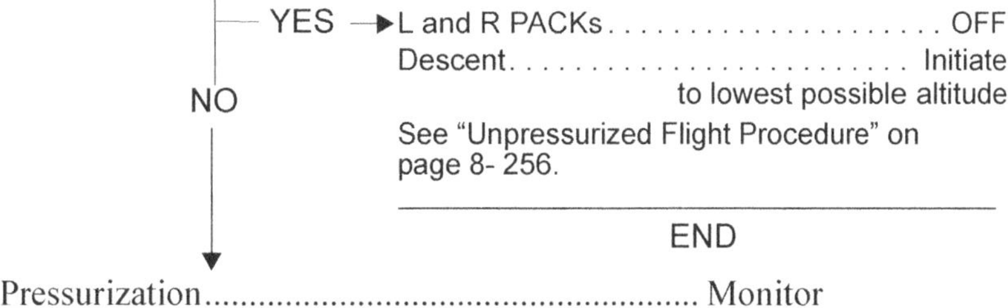

Pressurization... Monitor

L (R) PACK TEMP

Choose a scenario.

AUTO mode:

Affected PACK .... OFF

NOTE:
Airplane altitude maximum 25,000 feet during single pack operations.

Pressurization .... Monitor

Does the PACK TEMP caution message persist?

YES → BLEED SOURCE .... Select alternate source to isolate affected pack

Notes:
1. Do not select the APU as the bleed source if wing or cowl anti-ice is required.
2. APU bleed can only be used to a maximum of 25,000 feet for ECS.

ISOL . . . . CLSD
BLEED VALVES . . . . MANUAL
WING A/I CROSS BLEED. . . . . Select non-affected side

ANTI-ICE, LH or RH COWL . . Affected side off
Leave icing conditions to prevent ice accumulation on inoperative cowl.

Note: Icing conditions exist in-flight at a TAT of 10°C (50°F) or below, and visible moisture in any form is encountered (such as clouds, rain, snow, sleet or ice crystals), except when the SAT is -40°C (-40°F) or below.

END

NO ↓

No further action required.

CONTINUED ON NEXT PAGE

MANUAL mode:

Pack discharge temperature is <5°C or >85°C

Affected Manual Mode

Temperature Control.................................. As required to maintain pack discharge between 5°C and 85°C.

Affected PACK discharge temp. .... Monitor

CONTINUED ON NEXT PAGE

Does the PACK TEMP caution message persist

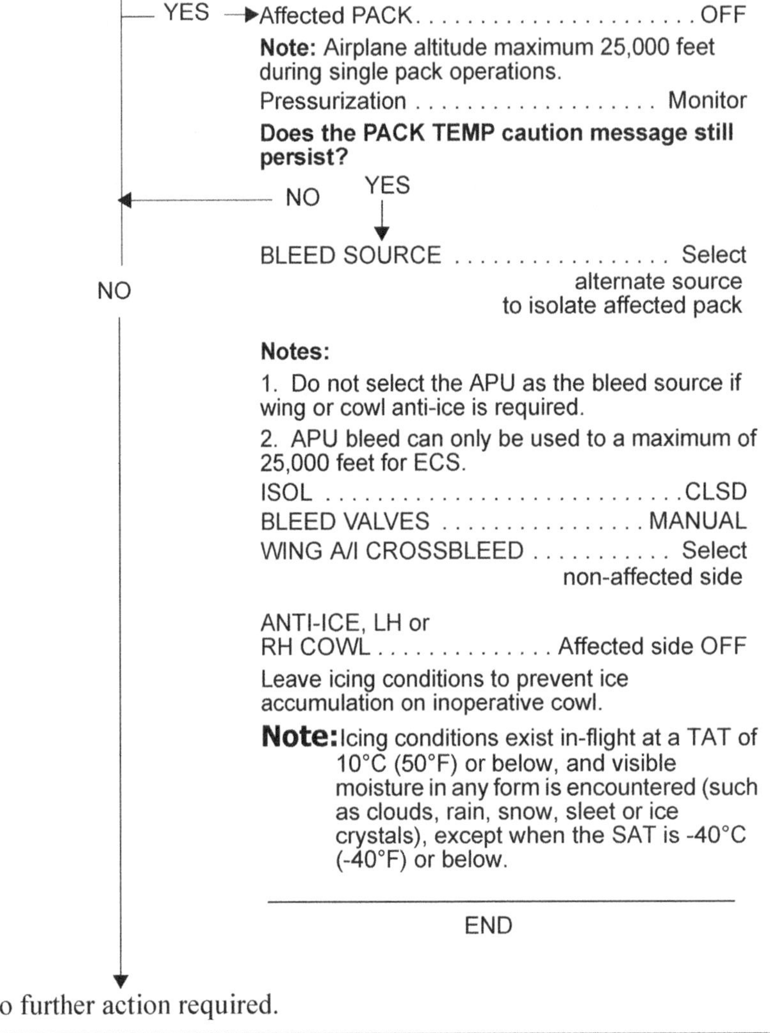

YES → Affected PACK . . . . OFF

Note: Airplane altitude maximum 25,000 feet during single pack operations.

Pressurization . . . . Monitor

Does the PACK TEMP caution message still persist?

NO (go to "No further action required.") YES ↓

BLEED SOURCE . . . . Select alternate source to isolate affected pack

Notes:

1. Do not select the APU as the bleed source if wing or cowl anti-ice is required.
2. APU bleed can only be used to a maximum of 25,000 feet for ECS.

ISOL . . . . CLSD

BLEED VALVES . . . . MANUAL

WING A/I CROSSBLEED . . . . Select non-affected side

ANTI-ICE, LH or RH COWL . . . . Affected side OFF

Leave icing conditions to prevent ice accumulation on inoperative cowl.

Note: Icing conditions exist in-flight at a TAT of 10°C (50°F) or below, and visible moisture in any form is encountered (such as clouds, rain, snow, sleet or ice crystals), except when the SAT is -40°C (-40°F) or below.

END

NO ↓

No further action required.

L (R) PACK AUTOFAIL

Affected AIR-CONDITIONING MAN.................. MAN

Affected Manual Mode

Temperature Control......................................As required

L (R) PACK

Affected PACK... OFF

NOTE:
Airplane altitude maximum 25,000 feet during single pack operations.

Pressurization... Monitor

Does the PACK caution message persist?

— YES → BLEED SOURCE . . . . Select alternate source to isolate affected pack

Notes:

1. APU bleed can only be used to a maximum of 25,000 feet for ECS.
2. Do not select the APU as the bleed source if wing or cowl anti-ice is required.

ISOL . . . . CLSD

BLEED VALVES . . . . MANUAL

WING A/I CROSS BLEED . . . . Select non-affected side

ANTI-ICE, LH
or RH COWL. . . . . Affected side OFF

Leave icing conditions to prevent ice accumulation on inoperative cowl.

Note: Icing conditions exist in-flight at a TAT of 10°C (50°F) or below, and visible moisture in any form is encountered (such as clouds, rain, snow, sleet or ice crystals), except when the SAT is -40°C (-40°F) or below.

END

NO ↓

No further action required.

AFT CARGO OVHT

AIR-CONDITIONING, AFT CARGO .... AIR

Does the CARGO OVHT caution message persist?

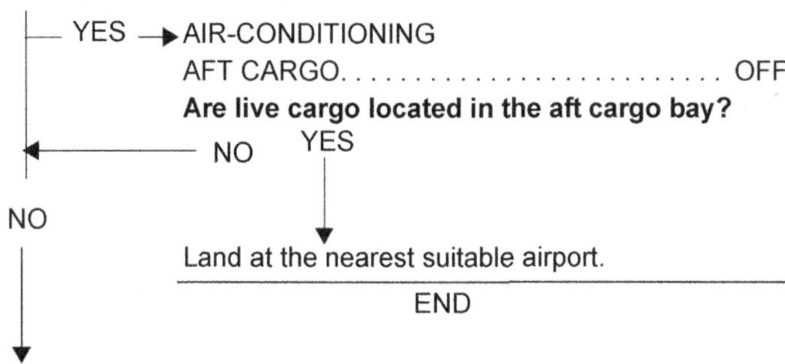

No further action required.

AVIONICS FAN

Choose a scenario.

In flight:

Avionics Fan ....FLT ALTN

On the ground:

Do not takeoff.

OVBD COOL

On the ground only:

Do not takeoff.

(Consider fault reset attempt. See page 4-26.)

DISPLAY COOL

If the AVIONICS FAN caution message is also displayed, action the AVIONICS FAN caution message first.

Choose a scenario.

In flight:

DISPLAY FAN ....FLT ALTN

After 60 seconds, does the DISPLAY COOL caution message persist?

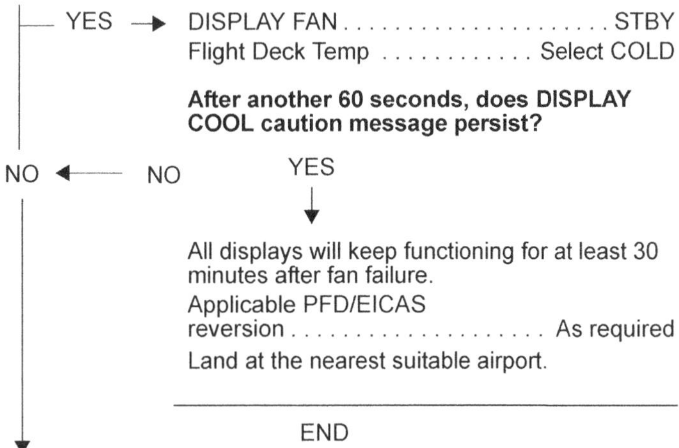
YES → DISPLAY FAN .... STBY
Flight Deck Temp .... Select COLD

After another 60 seconds, does DISPLAY COOL caution message persist?

NO ← NO

YES ↓

All displays will keep functioning for at least 30 minutes after fan failure.

Applicable PFD/EICAS reversion .... As required

Land at the nearest suitable airport.

END

No further action required.

CONTINUED ON NEXT PAGE

On the ground:

DISPLAY FAN... GND ALTN

After 60 seconds, does the DISPLAY COOL caution message persist?

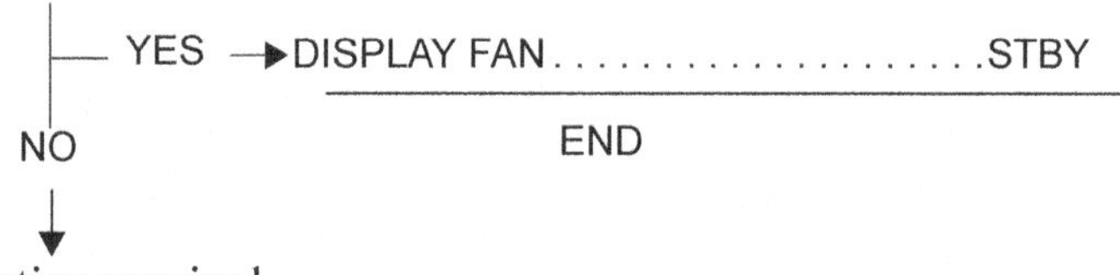

No further action required.

BLEED MISCONFIG

BLEED VALVES. . . . . .AUTO

NOTE:
If AUTO is inoperative, it may be necessary to cycle the BLEED VALVES (select from MANUAL to CLSD) to allow a reconfiguration of the bleed system.

Is manual bleed operation required?

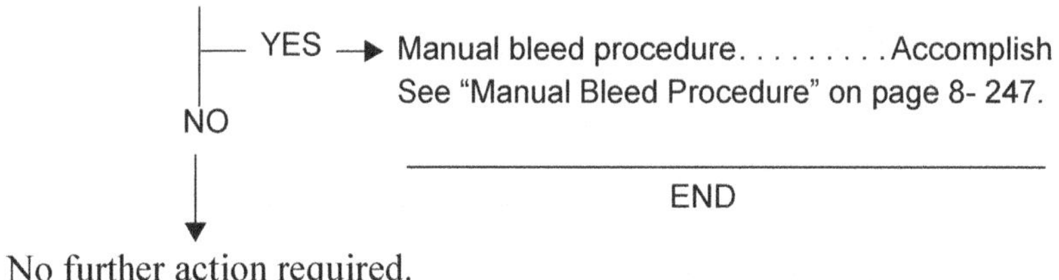

No further action required.

Manual Bleed Procedure

Choose a scenario.

To select both engines as the bleed source:

NOTE:
Operate only one pack per engine.

ISOL .... CLSD

BLEED SOURCE.. BOTH ENG

BLEED VALVES...MANUAL

To select only one engine as the bleed source:

NOTE:
Operate only one pack per engine.

ISOL .... CLSD

NOTE:
If it is necessary to operate a pack from the opposite engine bleed, select the ISOL to OPEN.

BLEED SOURCE..L or R ENG

BLEED VALVES...MANUAL

Inoperative PACK .... OFF

NOTE:
Airplane altitude maximum 25,000 feet during single pack operations.

WING A/I CROSS BLEED.....................................Select source engine side

ANTI-ICE, LH or RH COWL .... Non-source engine side OFF

Leave icing conditions to prevent ice accumulation on inoperative cowl.

CONTINUED ON NEXT PAGE

NOTE:
Icing conditions exist in-flight at a TAT of 10°C (50°F) or below, and visible moisture in any form is encountered (such as clouds, rain, snow, sleet or ice crystals), except when the SAT is -40°C (-40°F) or below.

To select the APU as the bleed source:

NOTE:
Do not select the APU as the bleed source if wing or cowl anti-ice is required.

NOTE:
APU bleed can only be used to a maximum of 25,000 feet for ECS

APU (37,000 feet and below) .... Start

ANTI-ICE, WING .... OFF

ANTI-ICE, LH and RH COWL................................. OFF

BLEED SOURCE selector .... APU

ISOL switch .... OPEN

BLEED VALVES selector .... MANUAL

To close engine bleeds and the APU LCV:

BLEED VALVES.. CLSD

NOTE:
The isolation valve will open when the BLEED VALVES is selected CLSD.

NOTE:
With the BLEED VALVES selected to CLSD, pressurization and wing and cowl anti-icing will be inoperative.

L (R) ENG BLEED

BLEED SOURCE selector .... Select non-affected engine source

ISOL switch .... CLSD

BLEED VALVES selector .... MANUAL

Affected side PACK switch .... OFF

NOTE:
Airplane altitude not above 25000 feet during single pack operations.

WING A/I CROSS BLEED selector .... Select non-affected side

ANTI-ICE, LH or RH COWL ....Affected side OFF

Leave icing conditions to prevent ice accumulation on inoperative cowl.

NOTE:
Icing conditions exist in-flight at a TAT of 10°C (50°F) or below, and visible moisture in any form is encountered (such as clouds, rain, snow, sleet or ice crystals), except when the SAT is -40°C (-40°F) or below.

(On the ground, if message remains, consider fault reset attempt. See page 4-26).

L (R) BLEED LOOP

BLEED SOURCE selector ....Select non-affected side

ISOL switch .... CLSD

BLEED VALVES selector ....MANUAL

WING A/I CROSS BLEED selector ....Select non-affected side

ANTI-ICE, LH or RH COWL ....Affected side OFF

Leave icing conditions to prevent ice accumulation on inoperative cowl.

NOTE:
Icing conditions exist in-flight at a TAT of 10°C (50°F) or below, and visible moisture in any form is encountered (such as clouds, rain, snow, sleet or ice crystals), except when the SAT is -40°C (-40°F) or below.

ISOL FAIL

Choose a scenario.

And the ISOL OPEN status message is shown:

BLEED SOURCE................................ L ENG or R ENG

ISOL ....OPEN

BLEED VALVES..MANUAL

Left or right PACK ....OFF

NOTE:
Airplane altitude maximum 25,000 feet during single pack operations.

WING A/I CROSS BLEED...........FROM OPERATIVE BLEED

ANTI-ICE, LH or RH COWL ....Affected side OFF

Leave icing conditions to prevent ice accumulation on inoperative cowl.

NOTE:
Icing conditions exist in-flight at a TAT of 10°C (50°F) or below, and visible moisture in any form is encountered (such as clouds, rain, snow, sleet or ice crystals), except when the SAT is -40°C (-40°F) or below.

And the ISOL CLSD status message is shown:

BLEED SOURCE... BOTH ENG

ISOL ....CLSD

BLEED VALVES..MANUAL

L (R) BLEED DUCT

CAUTION:

If l or r bleed duct warning message persists for 30 seconds, all bleed air sources will be closed causing loss of pressurization.

NOTE:

The L or R BLEED DUCT warning will be replaced by a L or R BLEED DUCT caution following automatic bleed valve closure and leak isolation.

ECS Synoptic Page ....Select

Are both L and R engine bleed valves closed?

YES → Descent ....Initiate immediately to 10,000 feet or lowest safe altitude whichever is higher

BLEED VALVES .... CLSD

Unpressurized Flight Procedure. . . Accomplish for PACKs OFF

See "Unpressurized Flight Procedure" on page 8-256.

Leave Icing Conditions

Note: Icing conditions exist in-flight at a TAT of 10°C (50°F) or below, and visible moisture in any form is encountered (such as clouds, rain, snow, sleet or ice crystals), except when the SAT is -40°C (-40°F) or below.

ANTI-ICE, WING .... OFF

ANTI-ICE, LH and RH COWL. . . . . OFF

END

NO ↓

Either L or R engine bleed valves are open:

BLEED SOURCE..Select

non-affected engine source

ISOL Valve..CLSD

BLEED VALVES...MANUAL

CONTINUED ON NEXT PAGE

Affected PACK .... Select Off

Altitude .... Not more than 25,000 feet

WING A/I CROSS BLEED..................................... Select

non-affected engine source

Affected COWL ANTI-ICE .... OFF

Leave Icing Conditions

NOTE:

Icing conditions exist in-flight at a TAT of 10°C (50°F) or below, and visible moisture in any form is encountered (such as clouds, rain, snow, sleet or ice crystals), except when the SAT is -40°C (-40°F) or below.

ALT LIMITER

Airplane altitude .... Not more than 25,000 feet

---

AUTO PRESS

MAN ALT .... HOLD

PRESS CONTROL... MAN

Manual pressurization control procedure ...... Accomplish

See "Manual Pressurization Control Procedure" on page 8- 255.

CABIN ALT (Amber)

BLEED VALVES .... AUTO

L and R PACK .... Confirm ON

EMER DEPRESS .... Confirm OFF

MAN ALT .... HOLD

PRESS CONTROL .... MAN, then RESET

Does the CABIN ALT caution message remain?

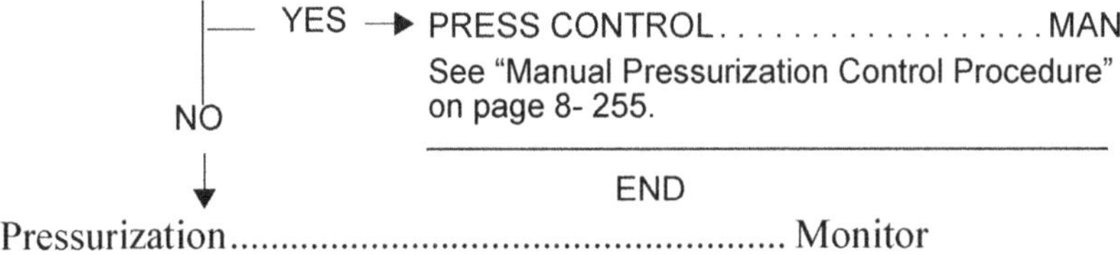

Pressurization .... Monitor

EMER DEPRESS

Is an emergency depressurization required?

EMER DEPRESS .... Confirm and select off

Manual Pressurization Control Procedure

MAN ALT .... HOLD

PRESS CONTROL.. MAN

To increase cabin altitude:

MAN ALT .... UP

MAN RATE..As required

To decrease cabin altitude:

MAN ALT .... DOWN

MAN RATE..As required

To maintain cabin altitude:

MAN ALT .... HOLD
when reaching target cabin altitude (see tables below)

Cruise FL	200	220	240	260	280	290
Target Cabin Alt	1500	2000	2400	2900	3500	3800

Cruise FL	310	330	350	370	390	410
Target Cabin Alt	4500	5300	6000	6700	7400	8000

Before landing:

Differential pressure .... Check not more than 1.0 PSID

NOTE:
Do not set cabin altitude below destination field elevation.

Is the differential pressure zero upon landing?

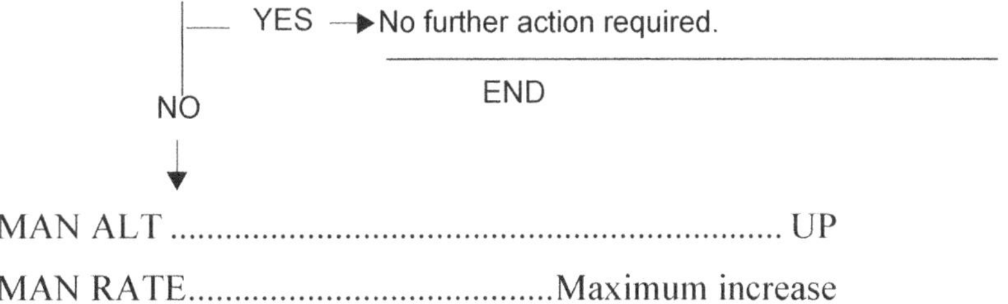

MAN ALT .... UP

MAN RATE.......................................Maximum increase

Unpressurized Flight Procedure

Choose a scenario

PACKs OFF:

Airplane Altitude .... Level at 10,000 feet maximum or lowest safe altitude, whichever is higher

Airspeed .... Not less than 210 KIAS

NOTE:
210 KIAS airspeed recommended during cruise to provide sufficient airflow to passengers within cabin.

NOTE:
When possible (terrain is not a factor) ensure aircraft altitude is below 13,700 feet prior to selecting the EMER DEPRESS switch to prevent oxygen mask deployment.

EMER DEPRESS switch .... Confirm and ON

L and R PACK switches .... OFF

RECIRC FAN switch.. OFF

AIR CONDITIONING, AFT CARGO switch .... OFF

RAM AIR switch ....Confirm and OPEN

PACKs ON:

Airplane Altitude .... 10,000 FEET MAXIMUM or lowest safe altitude

EMER DEPRESS .... Confirm and ON

Oxygen Index

Intentionally Left Blank

Oxygen

OXY LO PRESS

Is oxygen in use or is a leak suspected?

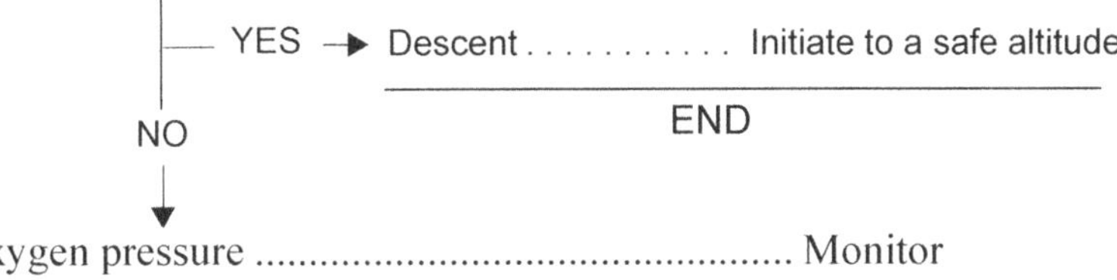

Oxygen pressure .... Monitor

Oxygen Pressure Readout Invalid (- - -)

Does the situation require the use of the crew oxygen masks?

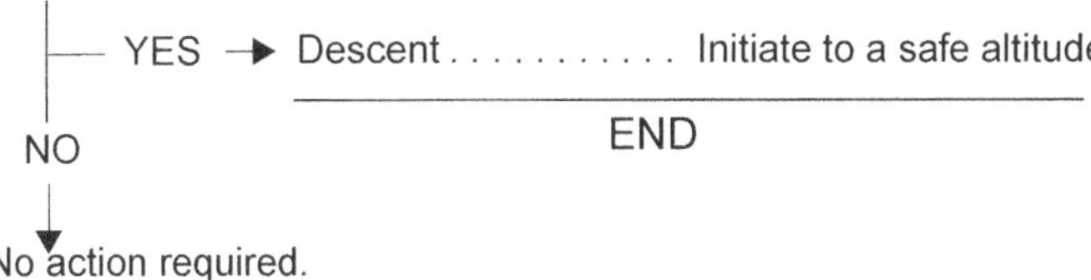

No action required.

PASS OXY ON

Passenger oxygen.. Check status

Passenger Oxygen - Auto Deploy Failure

Cabin altitude is above 14,250 (+/- 750 feet) and oxygen masks have not deployed automatically:

PASS OXY .... Confirm and ON

Did the masks deploy?

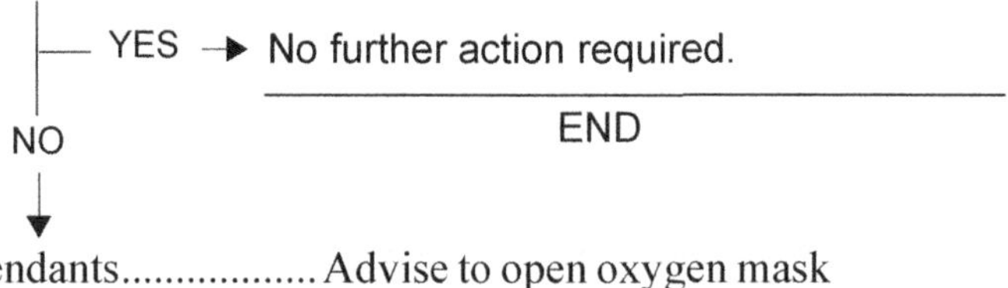

Flight attendants................. Advise to open oxygen mask compartments manually

Pilot Incapacitation

CEME: N1A

In case of pilot incapacitation, the following are general guidelines:

- Ensure the autopilot is engaged.
- Call the flight attendant to the flight deck.
- Check that the incapacited pilot does not interfere with the flight controls. It is preferable to have the incapacitated pilot removed from the flight deck.
- If unable to remove the incapacitated pilot, instruct flight attendant to lock the shoulder harness of the incapacitated pilot and move the seat rearwards.
- If an immediate landing is imperative, obtain advice on most suitable airport where medical assistance can be readily rendered.
- Check on the possibility of obtaining assistance from pilots who may be traveling as passengers on board the airplane.

Ice & Rain Protection Index

Intentionally Left Blank

Ice & Rain Protection

ICE

ANTI-ICE, WING ....ON

ANTI-ICE, LH and RH COWL...................................ON

WING OVHT

ANTI-ICE, WING .... OFF

Does the WING OVHT message persist after 40 seconds?

YES → Affected side. . . . . .Determine

BLEED SOURCE . . . . Select non-affected side

ISOL . . . . CLSD

BLEED VALVES . . . . MANUAL

WING A/I CROSS BLEED . . . . Select non-affected side

ANTI-ICE, LH or RH COWL . . . . Affected side OFF

ANTI-ICE, WING . . . . ON

Leave icing conditions to prevent ice accumulation on inoperative cowl.

Note: Icing conditions exist in-flight at a TAT of 10°C (50°F) or below, and visible moisture in any form is encountered (such as clouds, rain, snow, sleet or ice crystals), except when the SAT is -40°C (-40°F) or below.

END

NO ↓

ANTI-ICE, WING ....ON

ANTI-ICE DUCT

ANTI-ICE, WING ....OFF and Confirm
ANTI-ICE DUCT warning message out

Leave icing conditions.

Even small accumulations of ice on the wing leading edge can change the stall speed or stall characteristics of warning margins provided by the stall protection system.

NOTE:
Icing conditions exist in-flight at a TAT of 10°C (50°F) or below, and visible moisture in any form is encountered (such as clouds, rain, snow, sleet or ice crystals), except when the SAT is -40°C (-40°F) or below.

CONTINUED ON NEXT PAGE

Is ice accumulation observed on the heated portion of the wing leading edge?

YES → Airspeed .... Increase TO V_{MO}/M_{MO} to disperse ice, if possible.

Does ice accumulation persist?

YES → ANTI-ICE, WING .... ON

When ANTI-ICE DUCT warning msg comes on:

ANTI-ICE, WING .... OFF

Repeat above steps, as required (max 5 times).

Does ice accumulation on wing leading edge still persist?

YES → Maneuvering speed ..... Not less than 200 KIAS

Prior to landing:

GRND PROX, FLAP.........Confirm and OVRD

Landing flaps ....Use 20°

Appch Speed .... $V_{REF\ (FLAPS\ 45)}$ + 25 knots

Actual landing distance (with ice accumulation and flaps 20°) .... Increase

Without two Thrust Reversers and/or Wet/ Contaminated Runway Surface	With two Thrust Reversers and a Dry Runway Surface
1.50 (50%)	1.40 (40%)

Note: A slight pitch-up tendency may occur upon selection of reverse thrust. This can be readily corrected by the application of nose-down elevator and/or brakes.

END

NO

No further action required.

L (R) COWL A/I DUCT

ANTI-ICE, LH or RH COWL ....Affected side OFF

Leave icing conditions to prevent ice accumulation on inoperative cowl.

NOTE:

Icing conditions exist in-flight at a TAT of 10°C (50°) of below, and visible moisture in any form is encountered (such as clouds, rain, snow, sleet or ice crystals), except when the SAT is -40°C (-40°F) or below

Does the COWL A/I DUCT warning message persist?

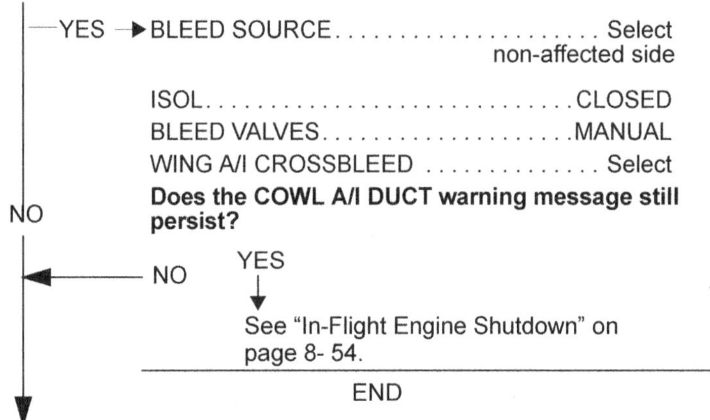

No further action required.

Ice Dispersal Procedure

After leaving icing conditions and if ice accumulation is observed on the heated portion of the wing leading edge:

Airspeed .... Increase TO V_{MO}/M_{MO}
if possible, to disperse ice

Is it still impossible to remove ice from the wing leading edge and assure an adequate stall margin?

YES → Maneuver speed .... Not less than 200 KIAS

WARNING: EVEN SMALL ACCUMULATIONS OF ICE ON THE WING LEADING EDGE CAN CHANGE THE STALL SPEED, STALL CHARACTERISTICS OR THE WARNING MARGINS PROVIDED BY THE STALL PROTECTION SYSTEM.

NO

Prior to landing:

GRND PROX, FLAP .... Confirm and OVRD

Landing Flaps .... Use 20°

Appch speed .... $V_{REF(FLAPS\ 45°)}$ + 25 KIAS

Actual landing distance
(with ice accumulation & flaps 20 ...... Increase

Without two Thrust Reversers and/or Wet/ Contaminated Runway Surface	With two Thrust Reversers and a Dry Runway Surface
1.50 (50%)	1.40 (40%)

Note: A slight pitch-up tendency may occur upon selection of reverse thrust. This can readily be corrected by the application of nose-down elevator and/or brakes.

Just prior to touchdown:

Thrust levers .... Retard to IDLE
and do not prolong the flare.

END

No further action required.

L (R) COWL A/I

NOTE:
With a COWL A/ICE caution message displayed, starter-assisted start is not available.

Is the Cowl Anti-Ice selected ON?

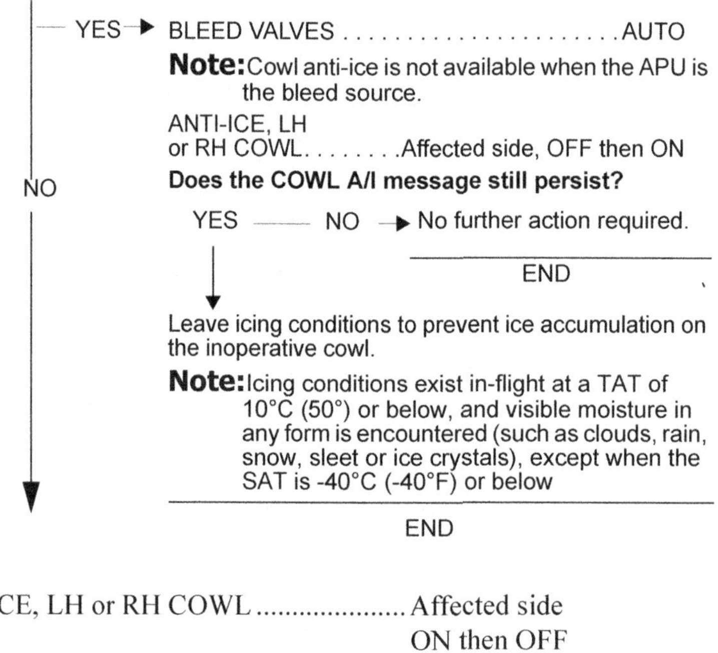

ANTI-ICE, LH or RH COWL .... Affected side ON then OFF

Engine instruments .... Monitor ITT

L (R) COWL A/I OPEN

NOTE:
With a COWL A/ICE OPEN caution message displayed, starter-assisted start is not available.

ANTI-ICE, LH or RH COWL .... Affected side ON then OFF

Engine instruments .... Monitor ITT

L (R) WING A/I

NOTE:
If L WING A/I and R WING A/I messages show, see "L and R WING A/I" on page 8-271.

This message indicates either a low temperature condition or a system failure.

Affected engine thrust......... Increase (75% N_2 minimum)

BLEED VALVES.. AUTO

Does the L or R WING A/I caution message persist?

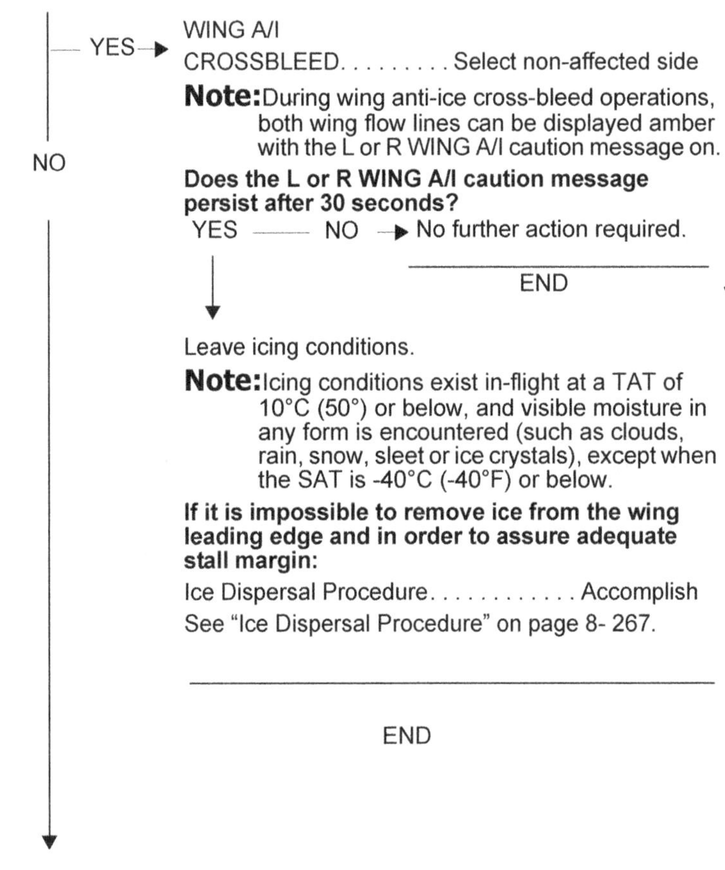

YES →

WING A/I
CROSSBLEED. . . . . Select non-affected side

Note: During wing anti-ice cross-bleed operations, both wing flow lines can be displayed amber with the L or R WING A/I caution message on.

Does the L or R WING A/I caution message persist after 30 seconds?

YES — NO → No further action required.

END

Leave icing conditions.

Note: Icing conditions exist in-flight at a TAT of 10°C (50°) or below, and visible moisture in any form is encountered (such as clouds, rain, snow, sleet or ice crystals), except when the SAT is -40°C (-40°F) or below.

If it is impossible to remove ice from the wing leading edge and in order to assure adequate stall margin:

Ice Dispersal Procedure. . . . . Accomplish

See "Ice Dispersal Procedure" on page 8- 267.

END

NO ↓

No further action required.

L and R WING A/I

This message indicates either a low temperature condition or a system failure.

Engine thrust....................... Increase (75% N_2 minimum)

BLEED VALVES.. AUTO

Do the L and R WING A/I caution messages persist?

YES → Leave icing conditions

Note: Icing conditions exist in-flight at a TAT of 10°C (50°) or below, and visible moisture in any form is encountered (such as clouds, rain, snow, sleet or ice crystals), except when the SAT is -40°C (-40°F) or below.

If it is impossible to remove ice from the wing leading edge and in order to assure adequate stall margin:

Ice Dispersal Procedure . . . . Accomplish

See "Ice Dispersal Procedure" on page 8- 267.

END

NO ↓

No further action required.

ANTI-ICE LOOP

On the ground only:

Do not takeoff.

L (R) COWL LOOP

On the ground only:

Do not takeoff.

WING A/I SNSR

ANTI-ICE, WING .... OFF

Leave icing conditions.

NOTE:
Icing conditions exist in-flight at a TAT of 10°C (50°) or below, and visible moisture in any form is encountered (such as clouds, rain, snow, sleet or ice crystals), except when the SAT is -40°C (-40°F) or below.

If it is impossible to remove ice from the wing leading edge and in order to assure adequate stall margin:

Ice Dispersal Procedure .... Accomplish

See "Ice Dispersal Procedure" on page 8-267.

WING XBLEED

Is WING A/I CROSSBLEED set to NORM?

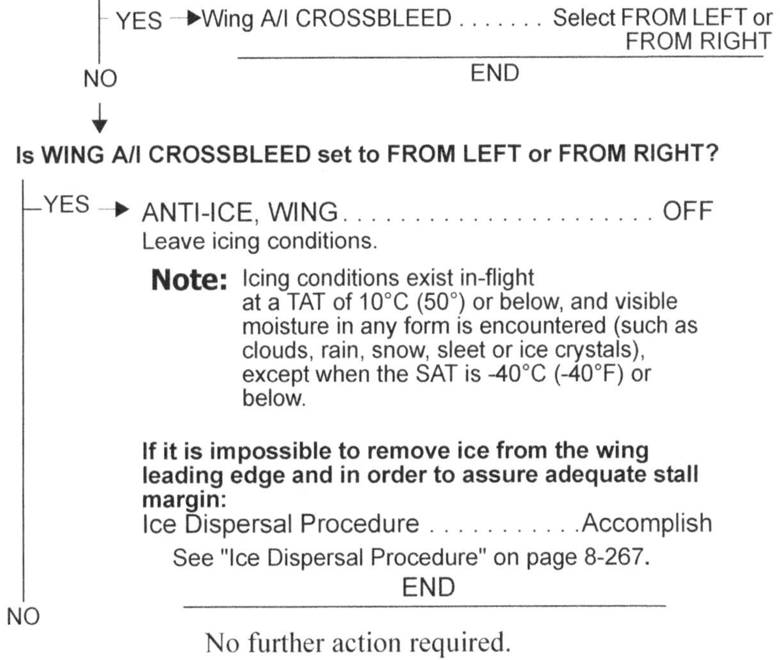

YES → Wing A/I CROSSBLEED .... Select FROM LEFT or FROM RIGHT

END

NO ↓

Is WING A/I CROSSBLEED set to FROM LEFT or FROM RIGHT?

YES → ANTI-ICE, WING .... OFF

Leave icing conditions.

Note: Icing conditions exist in-flight at a TAT of 10°C (50°) or below, and visible moisture in any form is encountered (such as clouds, rain, snow, sleet or ice crystals), except when the SAT is -40°C (-40°F) or below.

If it is impossible to remove ice from the wing leading edge and in order to assure adequate stall margin:

Ice Dispersal Procedure .... Accomplish

See "Ice Dispersal Procedure" on page 8-267.

END

NO

No further action required.

ICE DETECT FAIL

NOTE:
Icing conditions exist in-flight at a TAT of 10°C (50°) or below, and visible moisture in any form is encountered (such as clouds, rain, snow, sleet or ice crystals), except when the SAT is -40°C (-40°F) or below.

Are icing conditions present or anticipated?

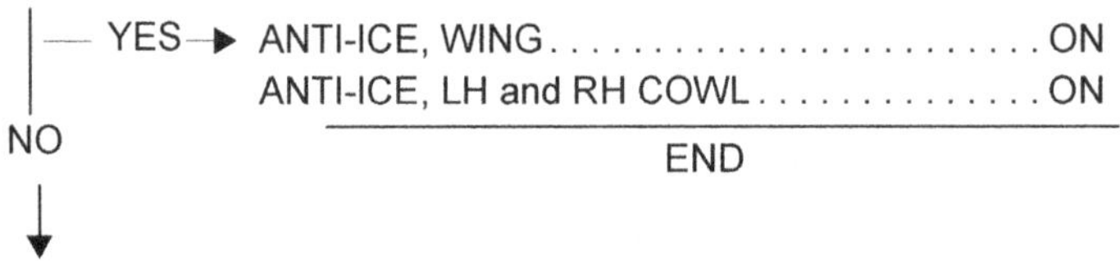

No further action required.

L and R AOA HEAT

ANTI-ICE, LH and RH PROBES .... OFF then ON

Do both L and R AOA HEAT caution messages persist?

YES ▶ Do not rely on stall protection and windshear recovery guidance systems.

STALL PTCT PUSHER. . . . . OFF
(either left or right)

Avoid icing conditions.

Note: Icing conditions exist in-flight at a TAT of 10°C (50°) of below, and visible moisture in any form is encountered (such as clouds, rain, snow, sleet or ice crystals), except when the SAT is -40°C (-40°F) or below.

Prior to landing:

Appch. speed . . . . $V_{REF\ (FLAPS\ 45°)}$ + 10 KIAS minimum

Actual landing distance. . . . . Increase

Without two Thrust Reversers and/or Wet/ Contaminated Runway Surface	With two Thrust Reversers and a Dry Runway Surface
1.15 (15%)	1.10 (10%)

END

NO

No further action required.

L (R) STATIC HEAT

ANTI-ICE, LH or RH PROBES .... Affected side
OFF then ON

Does the L (R) STATIC HEAT caution message persist?

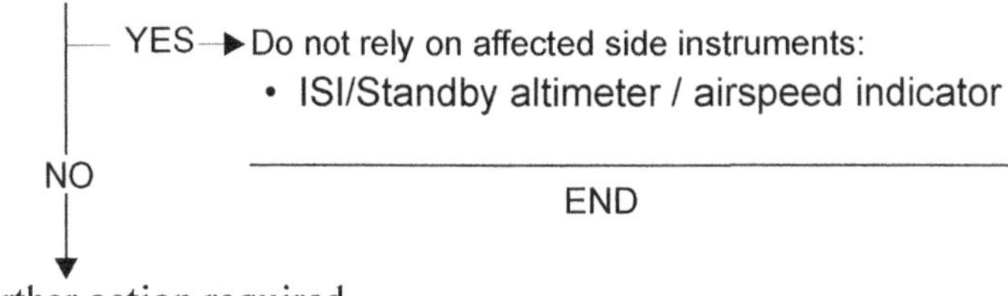

No further action required.

L (R) PITOT HEAT

ANTI-ICE, LH or RH PROBES .... Affected side
OFF then ON

Does the L (R) PITOT HEAT caution message persist?

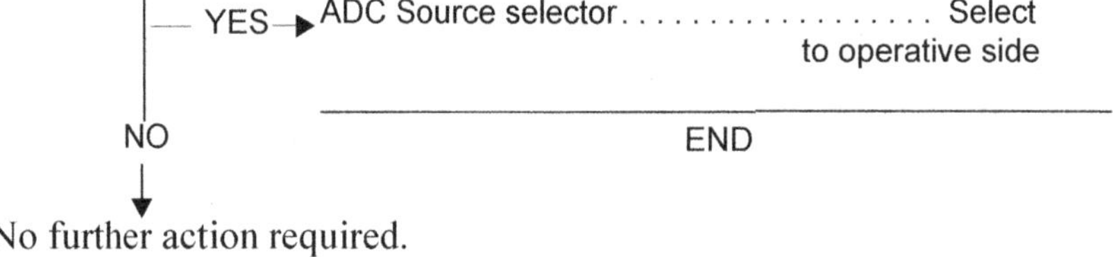

No further action required.

L (R) AOA HEAT

NOTE:
If L AOA HEAT and R AOA HEAT messages show, see "L and R AOA HEAT" on page 8-275.

ANTI-ICE, LH or RH PROBES.................. Affected side OFF then ON

Does the AOA HEAT caution message persist?

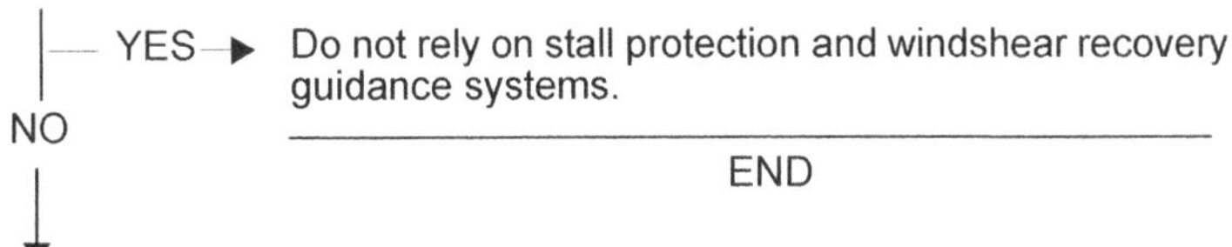

No further action required.

STBY PITOT HEAT

ANTI-ICE, LH PROBES............................ OFF then ON

Does the STBY PITOT HEAT caution message persist?

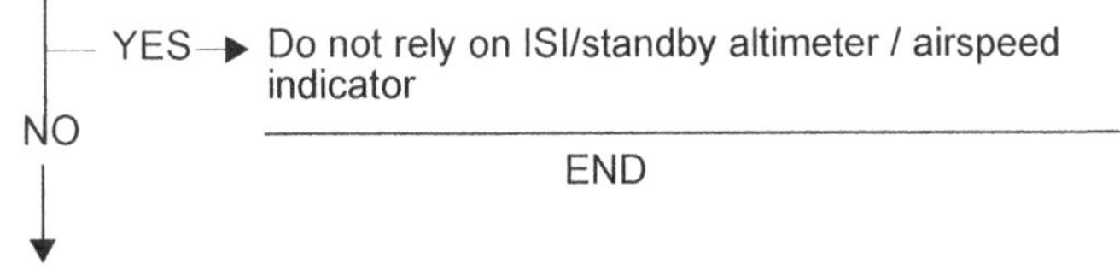

No further action required.

TAT PROBE HEAT

ANTI-ICE, LH PROBES............................ OFF then ON

Does the TAT PROBE HEAT caution message persist?

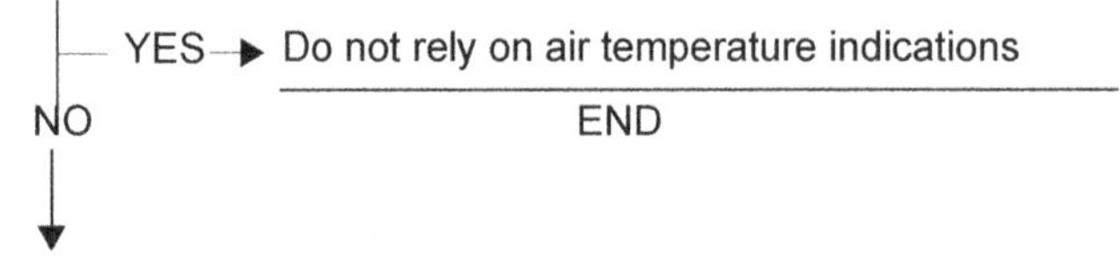

No further action required.

L (R) WSHLD HEAT

OR

L (R) WINDOW HEAT

ANTI-ICE, LH or RH WSHLD................... Affected side OFF/RESET then LOW or HI, as required.

Does the WSHLD HEAT or WINDOW HEAT caution message persist?

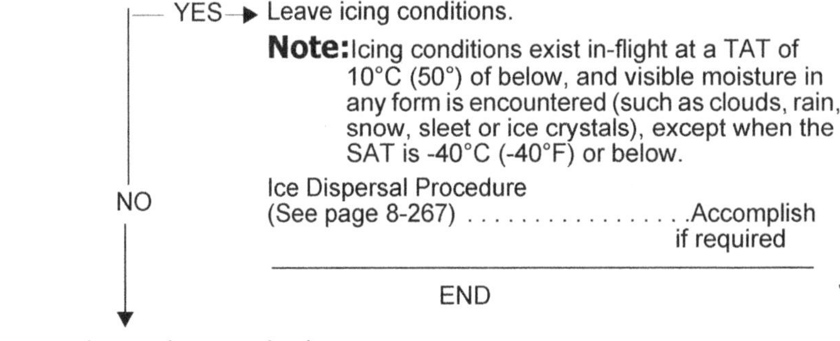

No further action required.

Window/Windshield (Arcing, Delaminated, Shatter, or Crack)

ANTI-ICE, LH or RH WSHLD..........Affected side, OFF

PRESS CONT.. MAN

MAN RATE selector .... INCR Max

MAN ALT ....Up (position), to achieve 6.2 psid or less

Crew and passenger oxygen .... Confirm and ON if required

Descent .... Initiate if required

Is the windshield core ply or inboard ply shattered?

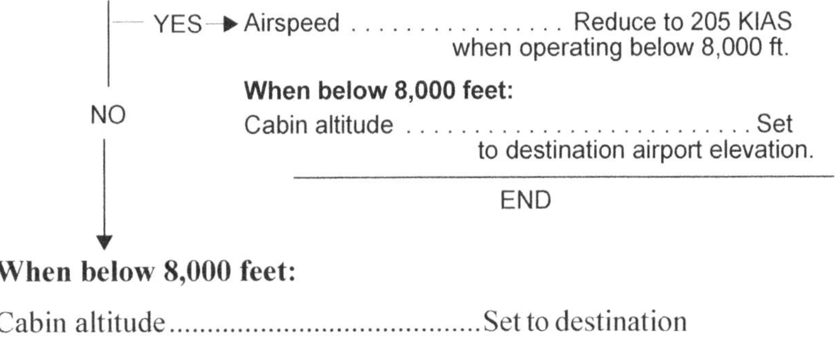
YES → Airspeed . . . . Reduce to 205 KIAS when operating below 8,000 ft.

When below 8,000 feet:

Cabin altitude . . . . Set to destination airport elevation.

END

NO ↓

When below 8,000 feet:

Cabin altitude...Set to destination airport elevation

Intentionally Left Blank

Landing Gear & Brakes Index

Intentionally Left Blank

Landing Gear & Brakes

NOTE:
If one or more landing gear symbols (nose, left or right main gear) display red hash marks (not safe), accomplish the applicable gear disagree procedure.

NOTE:
Failure to pull the LANDING GEAR MANUAL RELEASE handle to its full extension will result in one or more of the landing gear remaining up.

NOTE:
Landing gear warning horn comes on when landing gear is not extended with the airplane in the landing configuration.

BRAKE OVHT

IN FLIGHT:

Airspeed....................................Not more than 220 KIAS

LDG GEAR lever .... DN

BTMS indicators...................................Monitor until <15

BTMS OVHT WARN RESET................................Select

Is the BRAKE OVHT warning message out and all BTMS indicators displaying 06, or less?

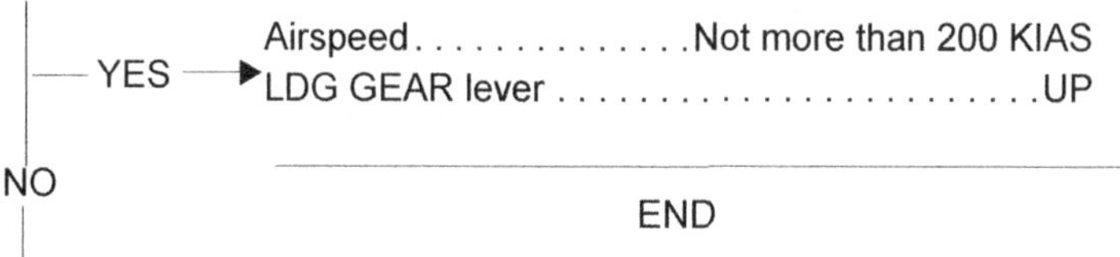

Land at the nearest suitable airport.

ON THE GROUND:

Brakes .... Minimize use

NOTE:
Use thrust reversers or shutdown one engine to minimize brake usage.

Do not takeoff.

CAUTION:
Suspect fire. there is a potential that the wheel fuse plug may release. keep personnel clear of the wheel bay area until the brake temperatures have cooled down.

PARKING BRAKE

PARKING BRAKE .... Check released

Does the PARKING BRAKE warning message persist?

YES → **Prior to landing:**

Note: The following landing distance factors are based upon both anti-skid systems inoperative.

Actual landing distance. . . . . Increase

Without two Thrust Reversers and/or Wet/ Contaminated Runway Surface	With two Thrust Reversers and a Dry Runway Surface
2.10 (110%)	1.75 (75%)

CAUTION: EXTREME CAUTION IS REQUIRED DURING BRAKING TO AVOID TIRE DAMAGE OR BLOWOUT. MAXIMIZE USE

END

NO ↓

No further action required.

During Landing - Excessive Asymmetry or Loss of Braking

Wheel brakes .... Release momentarily

ANTI SKID .... OFF

Wheel brakes .... Re-apply as required

CAUTION:

Extreme caution is required during braking to avoid tire damage or blowout. maximize use of reverse thrust.

GEAR DISAGREE

NOTE:
Confirm the LANDING GEAR MANUAL RELEASE handle is fully stowed.

Gear Up Disagree:

Airspeed....................................Not more than 200 KIAS

HYDRAULIC 3B pump..ON

LDG GEAR lever...DN

Land at the nearest suitable airport.

N/W STRG .... Select OFF then ARMD

NOTE:
Cycle the N/W STRG switch from ARMED to OFF and back to ARMED to enable nosewheel steering monitoring.

Does GEAR DISAGREE warning message persist?

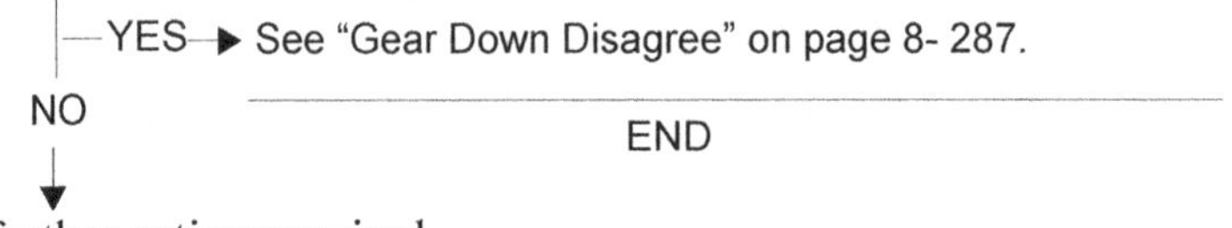

No further action required.

Gear Down Disagree

Airspeed .... Not more than 200 KIAS

HYDRAULIC 3B pump .... ON

LDG GEAR lever .... Select UP, then DN

NOTE:
If required, cycling of the LDG GEAR lever can be performed more than once.

Does the GEAR DISAGREE warning message persist?

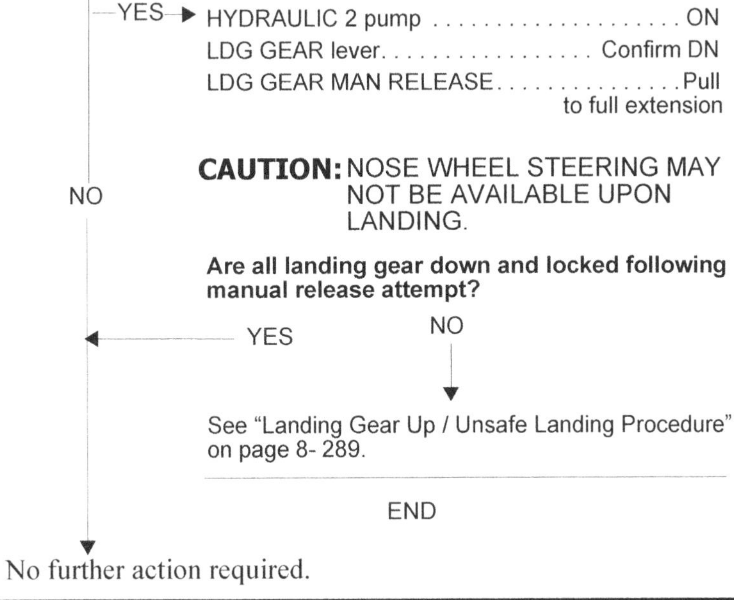

No further action required.

Landing Gear Lever Jammed in the UP Position

AirspeedNot more than 220 KIAS

HYDRAULIC 2 pump ....ON

LANDING GEAR
MANUAL RELEASE .... Pull to full extension

CAUTION:
Nose wheel steering will not be available upon landing.

NOTE:
The GEAR DISAGREE warning message will appear with the LDG GEAR lever jammed in the UP position and the landing gear down. Disregard the GEAR DISAGREE warning emergency procedure under this condition.

N/W STRG .... OFF

Are all landing gear down and locked following manual release attempt?

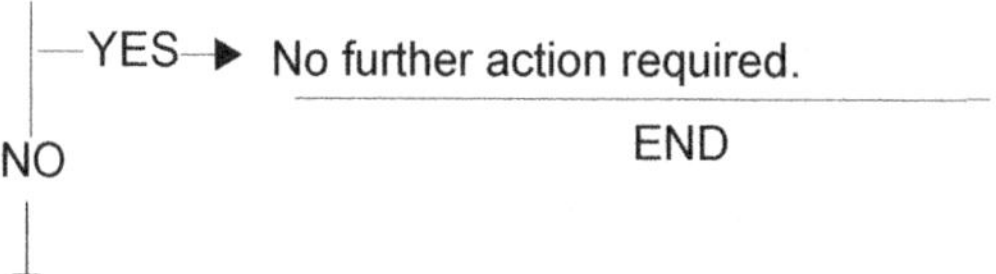

See "Landing Gear Up / Unsafe Landing Procedure" on page 8- 289.

Landing Gear Up / Unsafe Landing Procedure

PRELIMINARY:

Descent .... Plan

NOTE:
Reduce fuel to the minimum, if possible, while retaining sufficient fuel for a controlled, powered approach.

PREPARATION:

NOTE:
If one main landing gear is up or unsafe, hold applicable wing up as long as possible. Maintain directional control with rudder and nosewheel steering (if considered safe). When wing touches ground, apply asymmetrical braking for directional control.

NOTE:
If nose landing gear is up or unsafe, trim stabilizer nose-up after touchdown. Gently lower the nose before elevator effectiveness is lost.

NOTE:
If all wheels are up or unsafe, perform a nose high attitude touchdown, but do not reduce touchdown speed below stick shaker speed.

NOTE:
If both main landing gear cannot be locked down, consideration should be given to landing with all wheels up.

Crew and Flight Attendants .... Alert and brief

NOTE:
The briefing should include the following: Type of emergency, time available, airplane attitude after landing and exits available for use.

ATC .... Notify

NO PED and SEAT BLTS signs .... ON

Loose equipment .... Secure

CONTINUED ON NEXT PAGE

GRD PROX WARN cb (1B14) .... Open

AUDIO WARNING (All) .... DISABLE

NOTE:
Radio altitude callouts are not available.

Flight compartment door .... Unlocked

Shoulder harness .... Tight and locked

Plan to land with flaps 45°.

NOTE:
If 2 hydraulic systems failed, plan to land with FLAPS 20.

APPROACH:

L and R PACK .... OFF

PRESS CONT .... MAN

MAN ALT .... UP

When the airplane is completely depressurized:

BLEED VALVES .... CLSD

EMER LTS .... ON

At approximately 500 feet AGL:

Brace for impact .... Order over the PA system

BEFORE TOUCHDOWN:

APU FIRE PUSH .... Confirm and select

Airplane attitude .... Maintain nose high attitude

NOTE:
Ground/landing field contact should be accomplished using minimum forward speed, but not less than stick shaker speed, and at a minimum sink rate.

CONTINUED ON NEXT PAGE

AFTER LANDING:

Has landing gear collapsed, or failed to extend?

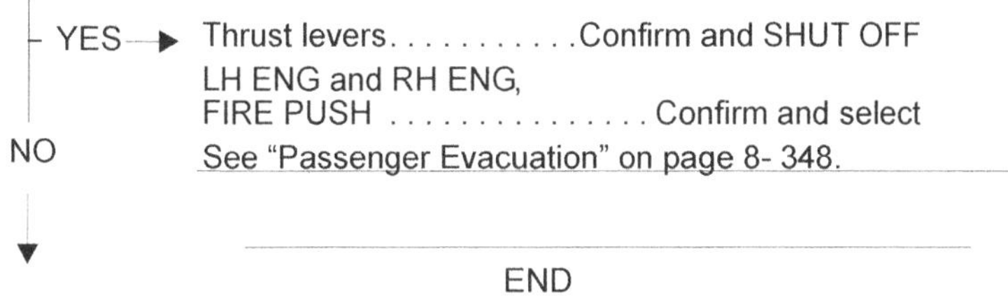

HYDRAULIC 2 and 3B pumps.......................... Both ON

Landing gear locking pins .... Install

MLG BAY OVHT

Airspeed.................................. Not more than 220 KIAS

LDG GEAR lever .... DN

BTMS indicators .... Monitor

Is MLG BAY OVHT warning message out and are BTMS indicators displaying 06, or less?

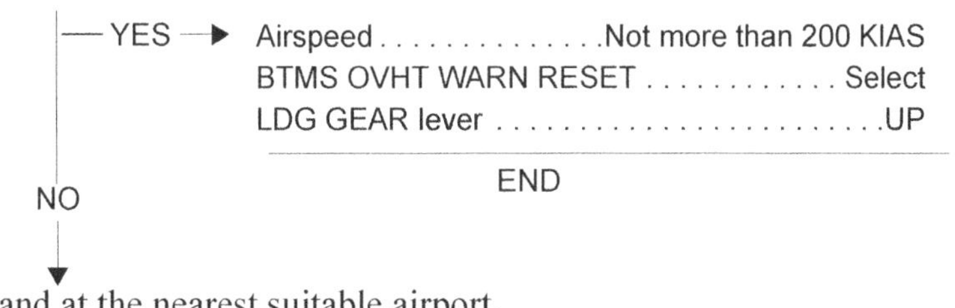

Land at the nearest suitable airport.

NOSE DOOR OPEN

Airspeed.................................. Not more than 220 KIAS

Land at the nearest suitable airport.

A/SKID INBD

NOTE:
If A/SKID OUTBOD also shows, see "A/SKID INBD and A/SKID OUTBD" on page 8-294.

NOTE:
Inboard brake BTMS indications are not available.

NOTE:
The landing distance factors below are based upon both anti-skid systems inoperative.

Actual landing distance....................................... Increase

Without two Thrust Reversers and/or Wet/ Contaminated Runway Surface	With two Thrust Reversers and a Dry Runway Surface
2.10 (110%)	1.75 (75%)

After touchdown:

FLIGHT SPOILER lever .... Select MAX deploy

CAUTION:
Extreme caution is required during braking to avoid tire damage or blowout. maximize use of reverse thrust.

A/SKID OUTBD

NOTE:
If A/SKID INBD also shows, see "A/SKID INBD and A/SKID OUTBD" on page 8-294.

NOTE:
Outboard brake BTMS indications are not available.

CONTINUED ON NEXT PAGE

NOTE:
The landing distance factors below are based upon both anti-skid systems inoperative.

Actual landing distance.. Increase

Without two Thrust Reversers and/or Wet/ Contaminated Runway Surface	With two Thrust Reversers and a Dry Runway Surface
2.10 (110%)	1.75 (75%)

After touchdown:

FLIGHT SPOILER lever.................. Select MAX deploy

CAUTION:
Extreme caution is required during braking to avoid tire damage or blowout. maximize use of reverse thrust.

A/SKID INBD and A/SKID OUTBD

NOTE:
Inboard and outboard brake BTMS indications are not available.

ANTI SKID..OFF

Actual landing distance..Increase

Without two Thrust Reversers and/or Wet/ Contaminated Runway Surface	With two Thrust Reversers and a Dry Runway Surface
2.10 (110%)	1.75 (75%)

NOTE:
Land with a firm touchdown to ensure that main gear weight-on-wheels signal is achieved for GLD deployment.

After touchdown:

FLIGHT SPOILER lever.................. Select MAX deploy

CAUTION:
Extreme caution is required during braking to avoid tire damage or blowout. maximize use of reverse thrust.

IB (OB) BRAKE PRESS

Hydraulic pressure and fluid quantity....... Check affected system

Brake Pressure .... Check

CONTINUED ON NEXT PAGE

Prior to landing:

Is the brake pressure less than 1800 psi for the applicable brake?

YES → Max landing weight . . . . Determine using the following max brake energy table and correct for wind and slope.

OAT		Airport Pressure Altitude (Feet)					
		0	2000	4000	6000	8000	10000
°C	**°F**	**Landing Weight (lbs.) Due to Max Brake Energy**					
-40	-40	85,980	85,980	85,980	85,980	84,244	80,743
-20	-4	85,980	85,980	85,980	84,181	80,766	77,456
0	32	85,980	85,980	84,156	80,870	77,684	74.542
20	68	85,980	84,302	81,058	77,922	74,898	71,927
40	104	84,597	81,388	78,289	75,293	72,343	69,477

Wind corrections:

Increase max landing weight by 2200 lbs per 10 kts headwind.

Decrease max landing weight by 8820 lbs per 10 kts tailwind

Runway Slope Corrections:

Increase max landing weight by 1540 lbs per 1% uphill slope

Decrease max landing weight by 1980 lbs per 1% downhill slope.

Note: The actual landing weight must not exceed the corrected maximum landing weight due to brake energy.

Actual landing distance . . . . Increase

Without two Thrust Reversers and/or Wet/ Contaminated Runway Surface	With two Thrust Reversers and a Dry Runway Surface
1.70 (70%)	1.50 (50%)

Note: With the inboard or outboard brakes inoperative, maximize the use of reverse

NO ↓

Use a steady brake application upon landing. Do not cycle the brakes.

PARK BRAKE SOV

PARKING BRAKE handle ....Fully down (released)

After landing:

Wheel chocks .... Install, before turning off hydraulic systems 2 and 3

NOTE:
The parking brake system will deplete more rapidly than normal with the parking brake set and hydraulic systems 2 and 3 selected off.

STEERING INOP

N/W STRG .... OFF

Use differential braking, rudder and engine thrust as required to assist directional control.

NOTE:
Select the longest runway available with minimum turbulence and crosswind.

NOTE:
In high crosswind conditions, rudder effectiveness may be limited after landing with maximum reverse thrust selected.

(On the ground, if message remains, consider fault reset attempt. See "Fault Reset Procedures" on page 4- 26.)

MLG OVHT FAIL

No action required. The MLG Bay overheat detection system is inoperative.

Landing Gear Manual Extension

Airspeed .... Not more than 220 KIAS

HYDRAULIC 3B pump .... ON

LDG GEAR lever .... DN

Did any landing gear fail to become down and locked?

— YES →

HYDRAULIC 2 pump. . . . . ON

LANDING GEAR MANUAL RELEASE. . . . . Pull to full extension

Note: Main landing gear extension relies upon free fall following a manual landing gear extension. This can take up to 40 seconds to accomplish.

WARNING: THE NOSE GEAR WILL NOT EXTEND DURING THE LANDING GEAR MANUAL EXTENSION PROCEDURE WITH 'HYD 2 LO PRESS" MESSAGE DISPLAYED. SIDE SLIP MAY BE REQUIRED FOR THE MAIN LANDING GEAR TO ACHIEVE DOWN LOCK.

CAUTION: NOSE WHEEL STEERING MAY BE INOPERATIVE.

If an unsafe landing gear condition persists:

Landing Gear Up/

Unsafe Landing Procedure . . . . Accomplish

See "Landing Gear Up / Unsafe Landing Procedure" on page 8- 289.

END

NO ↓

No further action required.

PROX SYSTEM

NOTE:
Nuisance "TERRAIN PULL UP" and "TOO LOW GEAR" aural warnings may be announced.

NOTE:
The following systems are inoperative:

- Landing gear indication and control
- Nose landing light
- Airplane door indications
- The no smoking and seat belts lights system

Prior to landing:

Airspeed....................................Not more than 220 KIAS

HYDRAULIC 2 and 3B pumps...................................ON

LDG GEAR lever ....DN

LDG GEAR MAN RELEASE ....Pull to full extension

N/W STRG ....OFF

NOTE:
Select the longest runway available with minimum turbulence and crosswind.

Actual landing distance.......................................Increase

Without two Thrust Reversers and/or Wet/ Contaminated Runway Surface	With two Thrust Reversers and a Dry Runway Surface
1.35 (35%)	1.30 (30%)

CONTINUED ON NEXT PAGE

CAUTION:
Touchdown protection for the brakes is lost. do not depress brake pedals until after touchdown.

After touchdown:

FLIGHT SPOILER lever......... Select MAX Deployment

NOTE:
Use differential braking, rudder and engine thrust as required to assist directional control.

At 30 knots (airplane speed):

NOTE:
This failure could lead to a loss of the inboard and outboard brakes during taxi. Revert to manual braking (anti-skid disabled) at low taxi speed.

ANTI SKID..OFF

After landing:

HEATERS circuit breakers..Open

- TAT cb (1A12)
- AOA R cb (1A13)
- PITOT R cb (1A14)
- STATIC R cb (1G14)
- PITOT L cb (1T7)
- AOA L cb (1T8)
- PITOT STBY cb (1T9)
- STATIC L cb (2S1)

CRT Display temperatures.................................. Monitor for possible overheat

CONTINUED ON NEXT PAGE

Before removing AC electrical power:

ADG DEPLOY AUTO cb (2N6)..............................Open

HYDRAULIC 3B pump..ON

Landing gear locking pins .... Install

NOTE:
Do not leave the APU unattended. Automatic APU fire extinguishing is not available on the ground.

PROX SYS CHAN

NOTE:
Nuisance "TERRAIN PULL UP" and "TOO LOW GEAR" aural warnings may be announced.

NOTE:
The nose landing light may be inoperative.

NOTE:
The no smoking (if installed) and seat belts lights system may be inoperative even if the NO SMOKING and SEAT BELTS status messages are displayed.

NOTE:
The landing gear UP symbol indication may be lost.

NOTE:
AIrplane door indications may be inoperative.

CAUTION:
Touchdown protection for the brakes is lost. do not depress brake pedals until after touchdown.

CONTINUED ON NEXT PAGE

After landing:

Before removing AC electrical power:

ADG DEPLOY AUTO cb (2N6)..............................Open

NOTE:
Do not leave the APU unattended. Automatic APU fire extinguishing is not available on the ground.

WOW INPUT

Prior to landing:

NOTE:
Nose wheel steering may be inoperative upon landing.

EMER DEPRESS .... Confirm and ON

Actual landing distance.. Increase

Without two Thrust Reversers and/or Wet/ Contaminated Runway Surface	With two Thrust Reversers and a Dry Runway Surface
1.35 (35%)	1.30 (30%)

CAUTION:
Touchdown protection for the brakes is lost. do not depress brake pedals until after touchdown.

After touchdown:

FLIGHT SPOILER lever.......... Select MAX deployment

CONTINUED ON NEXT PAGE

At 30 knots (airplane speed):

NOTE:
This failure could lead to a loss of the inboard and outboard brakes during taxi. Revert to manual braking (anti-skid disabled) at low taxi speed.

ANTI SKID .... OFF

Prior to shutdown:

HEATERS circuit breakers....................................... Open

- TAT cb (1A12)
- AOA R cb (1A13)
- PITOT R cb (1A14)
- STATIC R cb (1G14)
- PITOT L cb (1T7)
- AOA L cb (1T8)
- PITOT STBY cb (1T9)
- STATIC L cb (2S1)

CRT Display temperatures .... Monitor
for possible overheat

Before removing AC electrical power:

ADG DEPLOY AUTO cb (2N6).............................. Open
to prevent inadvertent deployment
during APU/engine shutdown

NOTE:
Do not leave the APU unattended. Automatic APU fire extinguishing is not available on the ground.

WOW OUTPUT

NOTE:
Nose wheel steering may be inoperative upon landing.

Does the STALL FAIL caution message come on?

YES → STALL PTCT PUSHER

Left or right . . . . OFF

Appch. speed . . . . $V_{REF(flaps\ 45)}$+10 KIAS MIN.

Actual landing distance. . . . . Increase

Without two Thrust Reversers and/or Wet/ Contaminated Runway Surface	With two Thrust Reversers and a Dry Runway Surface
1.15 (15%)	1.10 (10%)

After landing, before removing AC electrical power:

ADG DEPLOY AUTO cb (2N6) . . . . .Open

END

NO ↓

No further action required.

Avionics/AFCS & WARN Systems Index

Intentionally Left Blank

Avionics/AFCS & WARN Systems

AFCS MSG FAIL or Autopilot Failure

Autopilot .... Disconnect
using AP/SP DISC switch on control wheel
or AP DISC switch on flight control panel.

NOTE:
Do not use autopilot when an AFCS MSG FAIL warning message is displayed.

NOTE:
Category II operations may be affected by this failure. Review the equipment requirements detailed in chapter 3 of this manual.

Primary Flight Display Failure

Affected display
reversionary panel selector .... Select PFD

NOTE:
Applicable MFD defaults to primary flight display.

NOTE:
Category II operations may be affected by this failure. Review the equipment requirements detailed in chapter 3 of this manual.

Display Control Panel Failure

DSPL CONT
Source Selector .... Select operative side

Autopilot XFR .... Select operative side

NOTE:
Category II operations may be affected by this failure. Review the equipment requirements detailed in chapter 3 of this manual.

NOTE:
HDG knob remains inoperative until operable display control panel is selected.

Air Data Computer Failure

NOTE:
RVSM operations are affected by an ADC failure. Review the equipment requirements found in chapter 4 of this manual.

Have both ADCs failed?

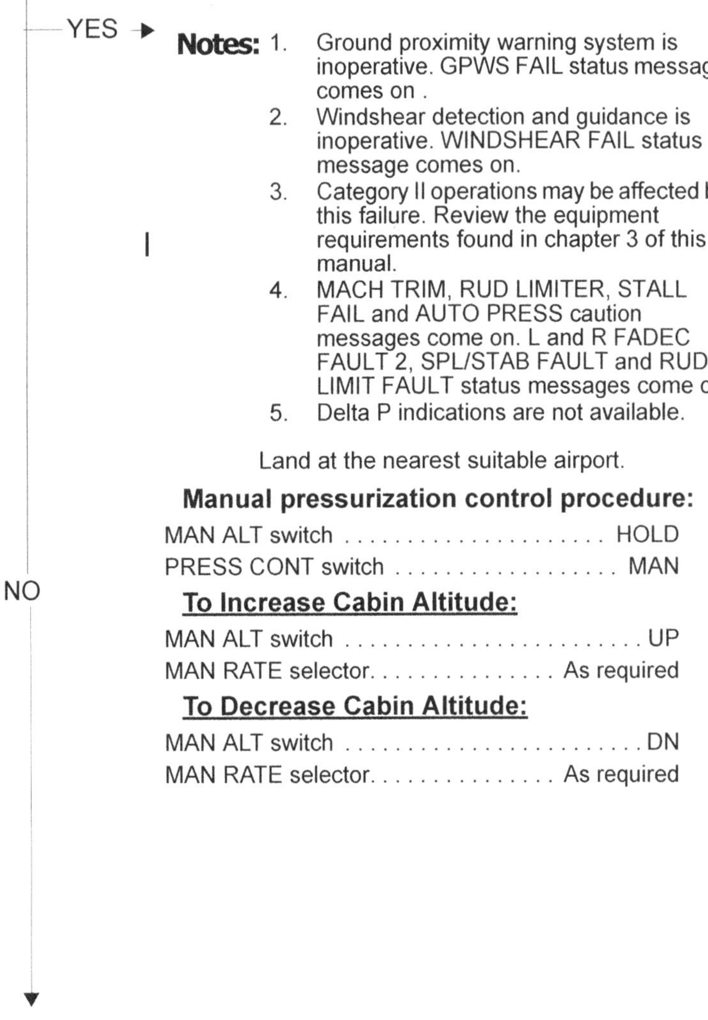

CONTINUED ON NEXT PAGE

To Maintain Cabin Altitude:

MAN ALT switch . . . . HOLD when reaching target cabin altitude (see table below)

Cruise FL	**180**	**200**	**220**	**240**	**260**	**280**
Target Cabin Alt.	1100	1500	2000	2400	2900	3500
Cruise FL	**290**	**310**	**330**	**350**	**370**	**390**
Target Cabin Alt.	3800	4500	5300	6000	6700	7400

NO

Before landing:

Cabin altitude . . . . Adjust to landing elevation

Note: Do not set cabin altitude below destination field elevation.

Upon touchdown:

MAN ALT switch . . . . UP

MAN RATE selector. . . . . Maximum INCR

END

CONTINUED ON NEXT PAGE

Has just the Captain's side ADC failed (ADC 1)?

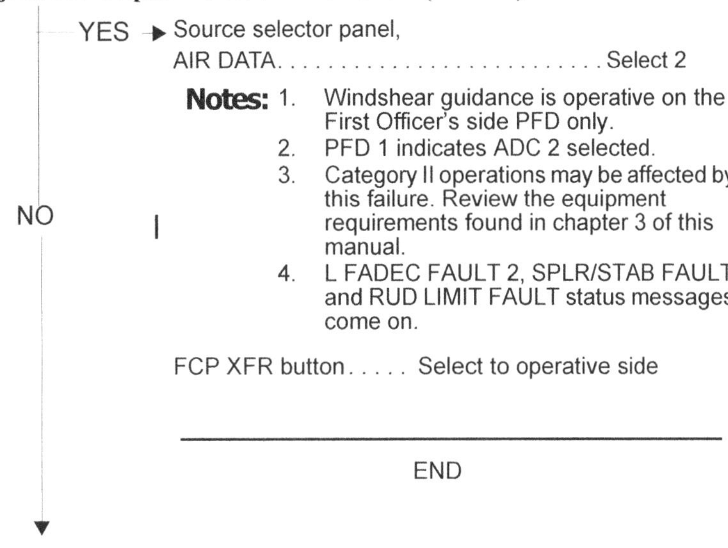

YES → Source selector panel,
AIR DATA. . . . . .Select 2

Notes:
1. Windshear guidance is operative on the First Officer's side PFD only.
2. PFD 1 indicates ADC 2 selected.
3. Category II operations may be affected by this failure. Review the equipment requirements found in chapter 3 of this manual.
4. L FADEC FAULT 2, SPLR/STAB FAULT and RUD LIMIT FAULT status messages come on.

FCP XFR button. . . . . Select to operative side

END

NO ↓

The First Officers ADC has failed (ADC 2):

Source selector panel,
AIR DATA. . . . . .Select 1

Notes:
1. Windshear guidance is operative on the Captain's side PFD only.
2. PFD 2 indicates ADC 1 selected.
3. Category II operations may be affected by this failure. Review the equipment requirements found in chapter 3 of this manual.
4. R FADEC FAULT 2, SPLR/STAB FAULT and RUD LIMIT FAULT status messages come on.

FCP XFR button. . . . . Select to operative side

END

DISPLAY TEMP Flag

(Any two displays indicate)

Land at nearest suitable airport

EFIS COMP INOP

Affected flight instruments .... Monitor

Standby flight instruments ....Cross-check

NOTE:
ISI ILS is valid for front course only.

EFIS COMP MON (IRS equipped Aircraft)

NOTE:
Category II operations may be affected by this failure. Review the equipment requirements as found in chapter 3 of this manual.

For IRS equipped aircraft, choose the appropriate PFD annunciation that is displayed along with the EFIS COMP MON caution message:

CONTINUED ON NEXT PAGE

HDG

EFIS and standby compass ....Cross-check

Autopilot .... Transfer to reliable side

Affected IRS .... Select ATT mode

FMS .... Enter present heading on IRS Control page

Is IRS operation in attitude mode acceptable?

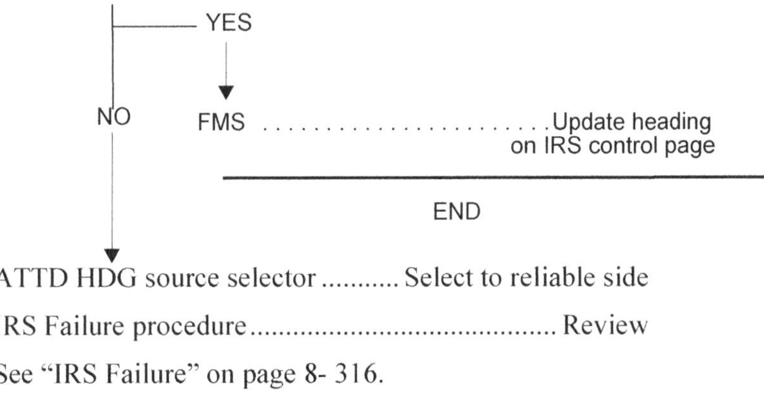

ATTD HDG source selector .... Select to reliable side

IRS Failure procedure.. Review

See “IRS Failure” on page 8- 316.

ROL or PIT

EFIS and ISI ....Cross-check

ATTD HDG source selector .... Select to reliable side

IRS Failure Procedure.. Review

See “IRS Failure” on page 8- 316.

CONTINUED ON NEXT PAGE

ALT or IAS

EFIS and ISI..Cross-check

AIR DATA source selector............ Select to reliable side

ADC Failure Procedure .... Review

See “Air Data Computer Failure” on page 8- 309.

FD ILS guidance is unreliable.

Radio Altimeter (1 or 2) Failure

NOTE:
Traffic alert and collision avoidance system is inoperative. TCAS FAIL status message comes on.

NOTE:
Ground proximity warning system is inoperative. GPWS FAIL status message comes on.

NOTE:
Windshear detection and guidance is inoperative. WINDSHEAR FAIL status message comes on.

NOTE:
Category II operations may be affected by this failure. Review the equipment requirements found in chapter 3 of this manual.

Have both RAs failed?

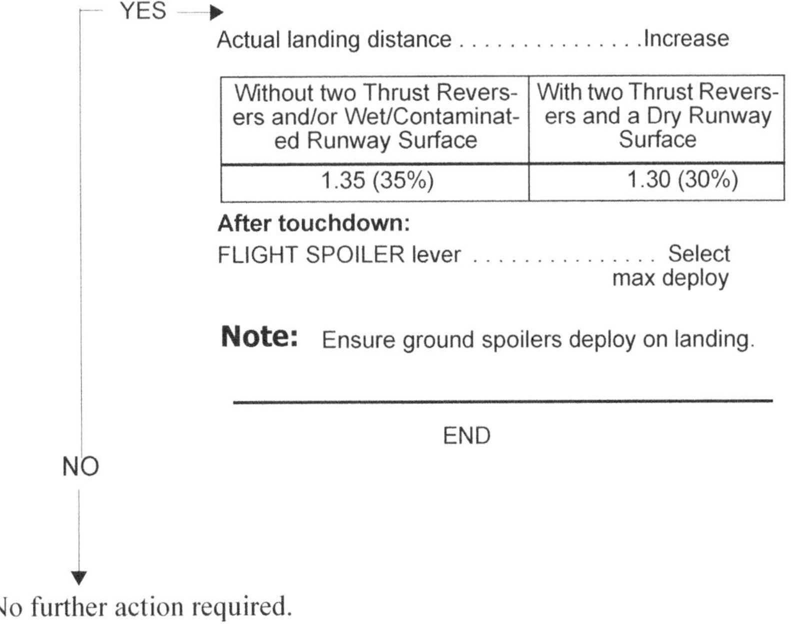

YES →

Actual landing distance . . . . Increase

Without two Thrust Reversers and/or Wet/Contaminated Runway Surface	With two Thrust Reversers and a Dry Runway Surface
1.35 (35%)	1.30 (30%)

After touchdown:

FLIGHT SPOILER lever . . . . Select max deploy

Note: Ensure ground spoilers deploy on landing.

END

NO ↓

No further action required.

CONTINUED ON NEXT PAGE

NOTE:
With one radio altimeter failed, both TCAS and GPWS remain operational.

NOTE:
If RA 1 has failed, windshear guidance is operative on PFD 2 only.

NOTE:
If RA 2 has failed, windshear guidance is operative on PFD 1 only.

IRS Failure

NOTE:
Autopilot and affected yaw damper will disengage.

NOTE:
Associated flight director is inoperative, and associated YD 1(2) INOP and FD 1(2) FAIL status messages come on.

NOTE:
Windshear guidance is operative on the non-affected PFD only.

NOTE:
Category II operations may be affected by this failure. Review the equipment requirements found in chapter 3 of this manual.

If IRS 1 has failed:

ATTD HDG source selector 2

Autopilot XFR switchTransfer to side with operative IRS

IRS 2 has failed:

ATTD HDG source selector 1

Autopilot XFR switchTransfer to side with operative IRS

Radio Tuning Unit Failure

Affected RTU INHIB switch..................................Select

Operable RTU..... Use 1/2 switch to tune cross-side radio

Have both RTUs failed?

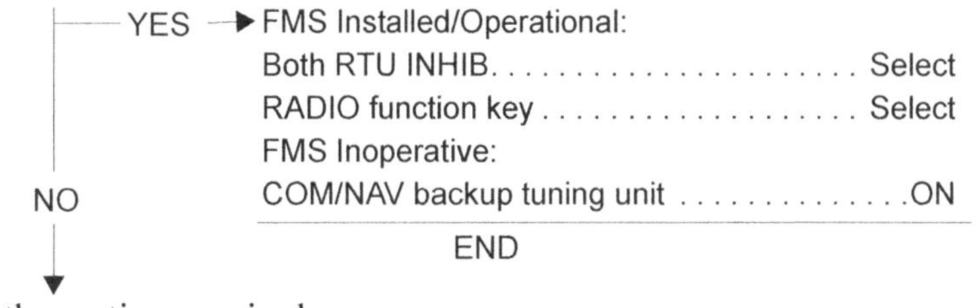

No further action required.

Altitude Reporting Transponder Failure

NOTE:
RVSM operations are affected by an altitude reporting transponder failure. Review the equipment requirements as found in chapter 4 of this manual.

Altitude Alerting System Failure

NOTE:
RVSM operations are affected by an altitude reporting transponder failure. Review the equipment requirements as found in chapter 4 of this manual.

AP TRIM IS LWD (RWD) or AP TRIM IS ND (NU))

NOTE:
Anticipate elevator out-of-trim condition when disconnecting the autopilot.

Autopilot .... Disconnect

Airplane .... Retrim

Autopilot .... Engage

Autopilot operation .... Monitor

Does the AP TRIM message persist?

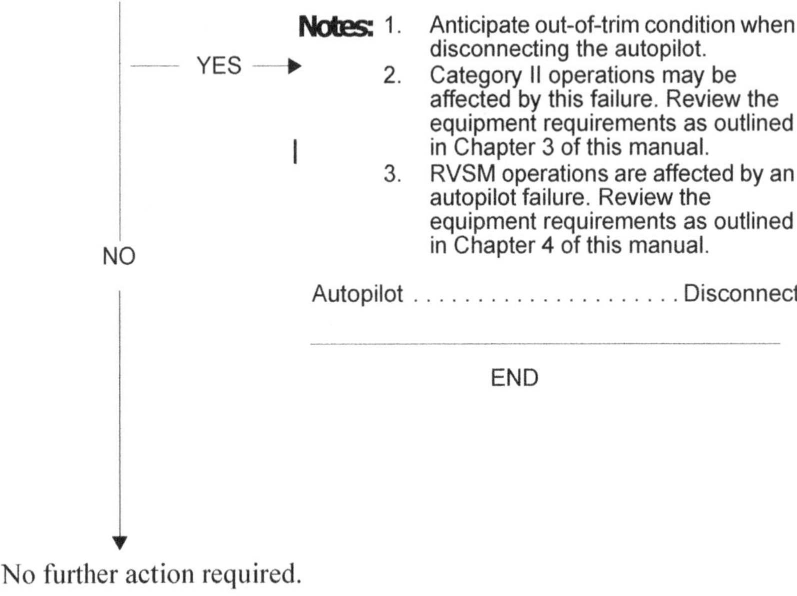

No further action required.

AP PITCH TRIM

NOTE:
Anticipate out-of-trim condition when disconnecting the autopilot.

Autopilot .... Disconnect

Airplane .... Retrim

Autopilot .... Engage

Autopilot operation .... Monitor

Does the AP PITCH TRIM message persist?

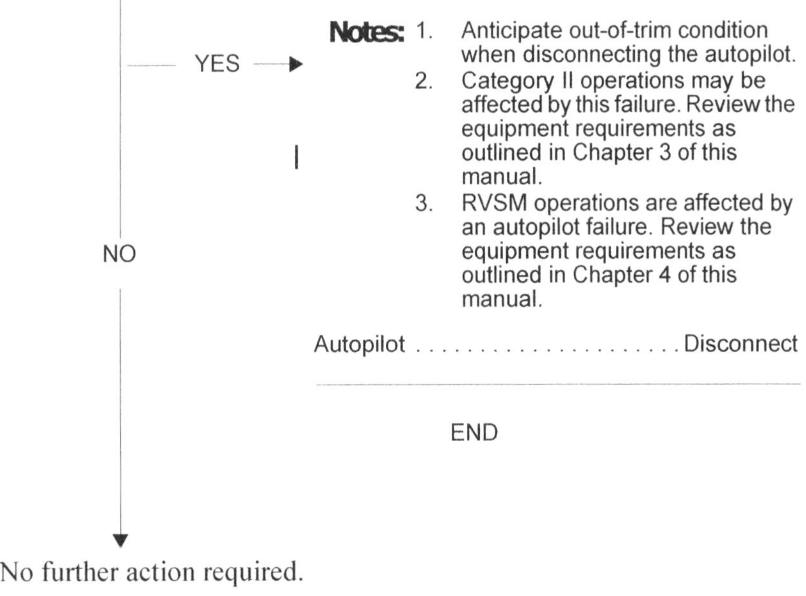

No further action required.

YAW DAMPER

YD1 and YD2 .... Engage

Does the YAW DAMPER caution message persist?

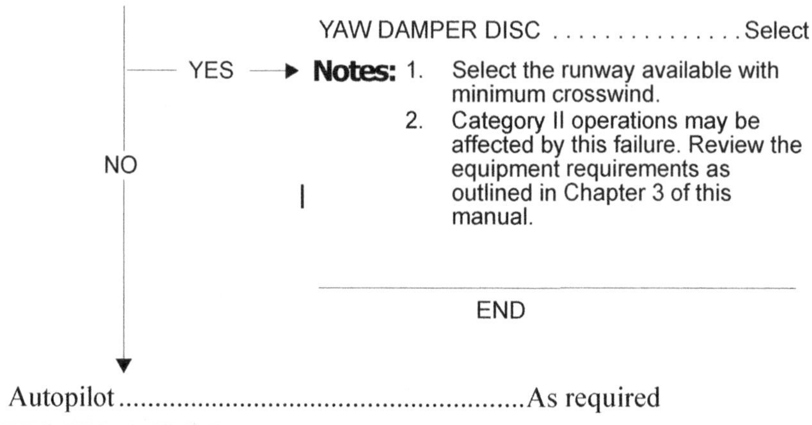

Autopilot .... As required

Flight Director Guidance Failed

NOTE:
Category II operations may be affected by this failure. Review the equipment requirements as outlined in Chapter 3 of this manual.

Choose a scenario

FD annunicator flag displayed on PFD:

AP XFER .... Select opposite side

NOTE:
Windshear guidance is operative on unaffected PFD only.

Sensor failure indicated (red line across affected indication):
Affected FD mode .... Deselect

Configuration Warning

Takeoff....................................Discontinue immediately

Check the position and status of the following:

- Aileron trim
- Autopilot
- Flaps
- Rudder trim
- Spoilers
- Stabilizer
- Parking Brake

NOTE:
The Memory item for Configuration Warning is designed for speeds below 80 KIAS. It is not recommended to discontinue takeoff solely based on Configuration Warnings that occur at speeds of 80 KIAS or greater unless supporting indications create the perception the aircraft is unsafe to fly. See CFM Chapter 7, "Rejected Takeoff (RTO)" for further information."

.

EICAS Primary Display Failure

EICAS source selector...ED 2

Applicable display reversionary panel selector.....EICAS

NOTE:
Applicable MFD defaults to EICAS STAT (status) page.

NOTE:
All page functions are selectable on the EICAS control panel.

EICAS Secondary Display Failure

Applicable MFD Display Selector.........................EICAS

NOTE:
Applicable MFD defaults to EICAS STAT (status) page.

NOTE:
All page functions are selectable on the EICAS control panel.

EICAS Control Panel Failure

To display a synoptic page (ECS, HYD, ELEC, FUEL, F/CTL, A/ICE, or DOORS:

STEP switch...............................Select STEP, as required

NOTE:
The PRI, STAT, CAS are operable.

NOTE:
ECS, HYD, ELEC, FUEL, F/CTL, A/ICE, DOORS, MENU, SEL, UP and DN are inoperative.

TCAS FAIL

(Message displayed on PFD/MFD)

Radio tuning unit, TCAS page

MODE line select key.. STBY

TCAS RA FAIL

(Message displayed on PFD)

Radio tuning unit, TCAS page

MODE line select key......................................TA ONLY

TCAS DISPLAY FAIL

(Message displayed on MFD)

Functional MFD.................................Use for TCAS data

---

DCU 1 (2) INOP or DCU 1 (2) AURAL INOP

(Shown as STATUS message)

Affected AUDIO WARNING .... Confirm and DISABLE

Operable AUDIO WARNING.............................. Normal

ELT ON

ELT switch................................ ON, then ARM / RESET

Check that ELT is not transmitting.

EMER LTS OFF

EMER LTS switch....................................... Check status of EMER LTS

AHRS Failure

NOTE:
Autopilot and affected yaw damper will disconnect.

NOTE:
Associated flight director is inoperative, associated YD 1(2) INOP and associated FD 1(2) FAIL status messages come on.

NOTE:
Windshear guidance is operative on non-affected PFD only.

NOTE:
Category II operations may be affected by this failure. Review the equipment requirements as outlined in Chapter 3 of this manual.

If AHRS 1 has failed:

Is the red MAG annunciation on the PFD?

YES → **If MAG mode is desired:**

ATTD HDG source selector 2

-or-

If DG mode is desired:

Affected COMPASS switch . . . . DG
slew heading to reliable side as often as required.

Autopilot XFR switch. . . . . Transfer to reliable side

END

NO ↓

CONTINUED ON NEXT PAGE

Any other failure annunciation is on the PFD:

ATTD HDG source selector 2

Autopilot XFR switch................. Transfer to reliable side

If AHRS 2 has failed:

Is the red MAG annunciation on the PFD?

YES → **If MAG mode is desired:**

ATTD HDG source selector 1

-or-

If DG mode is desired:

Affected COMPASS switch . . . . DG
slew heading to reliable side as often as required.

Autopilot XFR switch. . . . . Transfer to reliable side

END

NO

Any other failure annunciation is on the PFD:

ATTD HDG source selector 1

Autopilot XFR switch................. Transfer to reliable side

EFIS COMP MON (AHRS equipped Aircraft)

Indication: EFIS COMP MON caution message shown on AHRS equipped aircraft due to mismatch detected between on-side and cross-side data. At least one of the following comparator annunciators will be shown on the primary flight display.

- ALT, HDG, IAS, PIT, ROL, FD

NOTE:

Category II and RVSM operations may be affected by this failure. Review the equipment requirements as found in chapter 3 and 4 of this manual.

CONTINUED ON NEXT PAGE

For AHRS equipped aircraft, choose the appropriate procedure based on the scenario (on the ground or inflight) and PFD annunciation:

On The Ground:

HDG

NOTE:
The HDG flag and EFIS COMP MON message when shown on the ground may be caused by ground based equipment or other localized magnetic anomalies.

NOTE:
Known areas of magnetic field disturbances/anomalies are noted in the Jeppesen 10-7 pages.

NOTE:
NEVER attempt a takeoff when a HDG, MAG, or DG flag and/or EFIS COMP MON caution message is displayed.

COMPASS switches (both) .... Verify set to MAG

EFIS and standby instruments .... Cross-check to determine which AHRS has been affected

Affected AHRS COMPASS switch(es)........ Select to DG and then back to MAG.

Does the EFIS COMP MON caution message remain?

YES → If the aircraft is in an area of known or suspected magnetic anomalies, see "Takeoff From Runways with Known Magnetic Anomalies" on page 7-16.
If it is known that the aircraft has been moved away from all areas of magnetic anomalies, a system failure has likely occurred. DO NOT TAKEOFF.

NO ↓

No further pilot action required.

CONTINUED ON NEXT PAGE

In Flight:

HDG

EFIS and standby compass ....Cross-check

COMPASS.. Both MAG

AP XFER....................................... Select to reliable side

Affected COMPASS.........................Slew to reliable side

Does the EFIS COMP MON caution message reoccur?

YES → Affected COMPASS switch . . . . Select DG then MAG

NO ↓

Is DG mode desired (AP remains operational)?

YES → Affected COMPASS . . . . DG
Affected COMPASS . . . . Slew heading to reliable side as often as required.

END

NO ↓

If DG mode not acceptable (MAG mode desired):

ATTD HDG source selector . . . . Select to reliable side

AHRS Failure procedure . . . . Review
see AHRS Failure on page 8-323.

END

No further pilot action required.

CONTINUED ON NEXT PAGE

ROL or PIT

EFIS and standby flight instruments..............Cross-check

ATTD HDG source selector .... Select to reliable side

See "AHRS Failure" on page 8- 323.

ALT or IAS

EFIS and standby flight instruments..............Cross-check

AIR DATA source selector............. Select to reliable side

ADC Failure Procedure .... Review

See "Air Data Computer Failure" on page 8- 309.

No further pilot action required.

FD

FD ILS guidance is unreliable.

Uncommanded Yaw Motion

Indications
Abnormal / uncommanded change in yaw attitude
Continuous or intermittent rudder kicks
Sustained oscillation ("Dutch Roll")

Controls	Assume manual control and counter aircraft motion using handwheel inputs
YAW DAMPER, DISC button	Select to disconnect both yaw dampers

NOTE:
Disregard YAW DAMPER caution message.

Does the Uncommanded motion persist?

YES → Airspeed.....................Not more than 250 KIAS

Land at the nearest suitable airport.

NOTE: Select the runway available with minimum cross-wind and turbulence.

NO ↓

No further action required.

Intentionally Left Blank

Doors and Windows

PASSENGER DOOR

WARNING

Door failure may be indicated by loud noise, pressurization leak, or rumble emanating from the door area. if any of these indications are present, do not approach the door.

NO PED & SEAT BLTS ....Select both ON

Descent ....Initiate to 10,000 feet or lowest safe altitude.

NOTE:
Prepare to land at the nearest suitable airport.

NOTE:
Accomplish the "Rapid (Emergency) Descent" on page 7-90 if required.

PRESS CONTROL...MAN

MAN ALT ....UP

MAN RATE...MAX INCR

Land at the nearest suitable airport.

AV BAY DOOR or FWD CARGO DOOR
CTR CARGO DOOR or AFT CARGO DOOR
L (R) FWD EMER DOOR or FWD SERVICE DOOR
Crew Escape Hatch Unsafe or L (R) AFT EMER DOOR

SEAT BLTS ....ON

Cabin pressure...Check normal

Is cabin pressure normal?

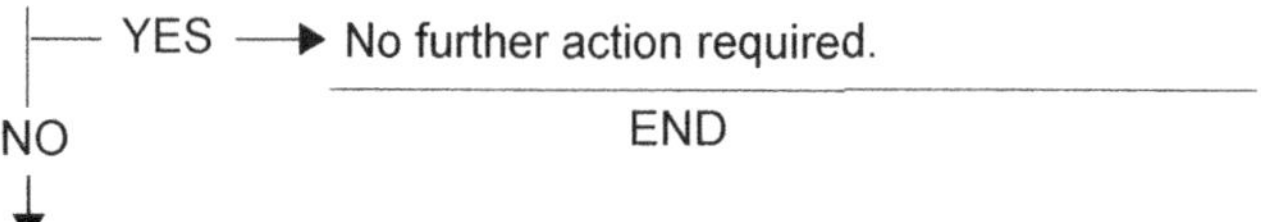

Switch to manual pressurization control:

MAN ALT .... HOLD

PRESS CONTROL.. MAN

MAN ALT ....As required

NOTE:
Maintain cabin altitude at 8,000 feet to minimize the pressure differential across the affected door.

NOTE:
Set the MAN ALT switch to "UP" to increase cabin altitude; "DN" to reduce cabin altitude.

MAN RATE...As required

commensurate with crew and pax comfort

When reaching 8,000 feet cabin altitude:

MAN ALT .... HOLD

NOTE:
Full cabin depressurization is not recommended.

CONTINUED ON NEXT PAGE

Descent .... Initiate
to 10,000 feet MSL or lowest
safe altitude, whichever is higher

Land at the nearest suitable airport.

At 10,000 feet or lowest safe altitude:

PRESS CONTROL.. AUTO

LDG ELEV.. SET
to landing field elevation

NOTE:
After landing, ensure that the airplane is completely depressurized prior to opening any airplane doors.

PAX DR OUT HNDL

SEAT BLTS sign..ON

Cabin pressure .... Check normal

Passenger Door .... Check visually

NOTE:
Check that green witness marks (registers) on all latches (rotary and sliding) are correctly aligned.

Is cabin pressure normal and green witness marks aligned?

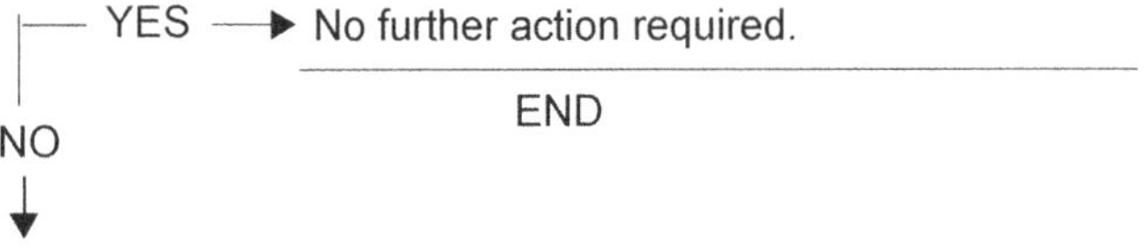

Switch to manual pressurization control:

MAN ALT .... HOLD

PRESS CONTROL... MAN

MAN ALT .... As required

CONTINUED ON NEXT PAGE

NOTE:
Maintain cabin altitude at 8,000 feet to minimize the pressure differential across the affected door.

NOTE:
Set the MAN ALT to "UP" to increase cabin altitude; "DN" to reduce cabin altitude.

MAN RATE...As required

commensurate with crew and pax comfort

When reaching 8,000 feet cabin altitude:

MAN ALT .... HOLD

Descent..Initiate
to 10,000 feet MSL or lowest
safe altitude, whichever is higher

Land at the nearest suitable airport.

At 10,000 feet or lowest safe altitude:

PRESS CONTROL. . . . . .AUTO
LDG ELEV . . . . Set
to landing field elevation

NOTE:
After landing ensure that the airplane is completely depressurized prior to opening any airplane doors.

PAX DR LATCH

SEAT BLTS sign ....ON

Cabin pressure ....Check normal

Is cabin pressure normal?

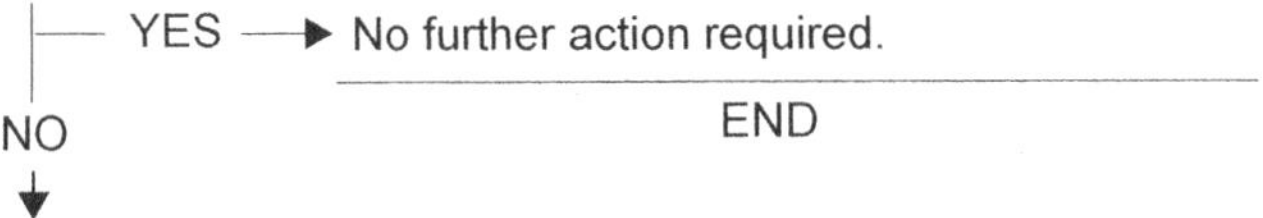

Switch to manual pressurization control.

MAN ALT .... HOLD

PRESS CONTROL.. MAN

MAN ALT ....As required

NOTE:
Maintain cabin altitude at 8,000 feet to minimize the pressure differential across the affected door.

NOTE:
Set the MAN ALT to "UP" to increase cabin altitude; "DN" to reduce cabin altitude.

MAN RATE..As required commensurate with crew and pax comfort

When reaching 8,000 feet cabin altitude:

MAN ALT .... HOLD

Descent ....Initiate to 10,000 feet MSL or lowest safe altitude, whichever is higher

Land at the nearest suitable airport.

CONTINUED ON NEXT PAGE

At 10,000 feet or lowest safe altitude:

PRESS CONTROL.. AUTO

LDG ELEV .... SET
to landing field elevation

NOTE:
After landing, ensure that the airplane is completely depressurized prior to opening any airplane doors.

Miscellaneous Emergency Procedures Index

Intentionally Left Blank

Miscellaneous Emergency Procedures

Rejected Takeoff

Simultaneously:

Thrust levers .... IDLE

Wheel brakes .... Maximum until safe stop

Thrust reverser(s) operating engine(s) .... Maximum consistent with directional control

After the airplane has been safely brought to a stop:

PARKING BRAKE .... On

Was takeoff rejected due to engine fire or severe damage?

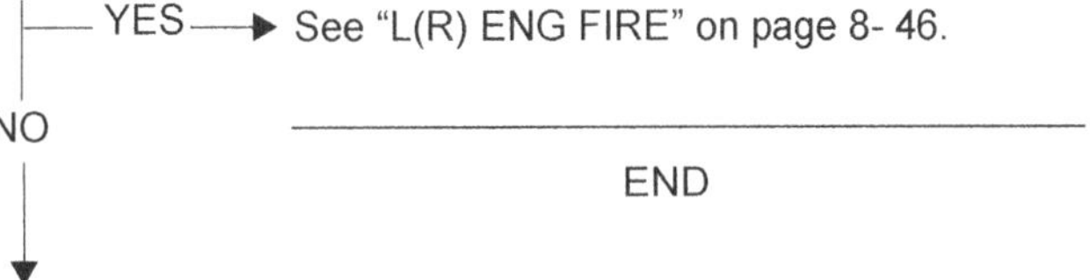

Is evacuation required?

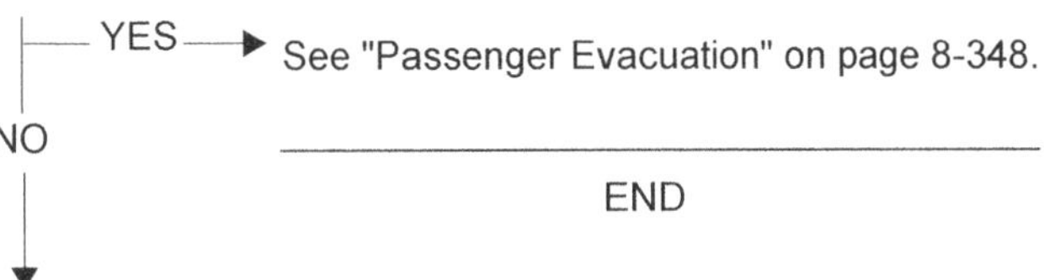

Passengers .... Advise to remain in their seats

CONTINUED ON NEXT PAGE

Post RTO Considerations

- Tower Notification: Once stopping is assured, notify ATC. Should a high energy (high speed, heavy weight) RTO occur, consider requesting emergency equipment be dispatched to meet the aircraft in the event of possible tire or brake fire.
- Airport Rescue and Fire Fighting (ARFF): ARFF personnel can provide important information regarding the exterior status of the aircraft. Consider remaining on the runway if emergency equipment is needed.
- RTO PA: If an evacuation is not required, make a PA as soon as practical and a second PA should be made after the situation has stabilized. If an evacuation is required, make a PA when directed by the Emergency Evacuation checklist. The crew must consider any hazard to the aircraft/ occupants (fire, cargo smoke) and take the appropriate action.
- Emergency/Non-Normal Checklists: Complete the RTO Non-Normal checklist.
- After Landing Checklist: Complete
- Prior to taxi: Check entry/exit door status to ensure an evacuation is not in progress. Ensure ground personnel and equipment are clear of the aircraft.
- Parking Brake: Consider releasing the parking brake, once chocked, to facilitate brake cooling and reduce the possibility of brakes fusing.
- Brake Cooling: Observe brake cooling schedule and precautions in this manual, starting on page 4-167.
- Subsequent Departures: Contact OCC.

NOTE:
After any RTO greater than 80 knots, the crew must seek approval to continue from a Chief Pilot, Fleet Manager or Director of Operations. The best method to facilitate this communication is through the OCC. All RTOs will be documented with a FSR.

Ditching and Forced Landing

Recommended configuration for water landing or forced landing.

Parameters	Water Landing	Forced Landing
Approach Speed	V_{REF}	V_{REF}
Descent Rate (if thrust is available)	200–300 fpm	200–300 fpm
Landing gear	Retracted	As required
Flaps Setting	45°	45°

Ditching or Forced Landing Imminent

L PACK and R PACK switches .... Select off

EMER DEPRESS switch .... Confirm and ON

Just before contact:

EMER DEPRESS switch .... Confirm and select off

Thrust levers .... Confirm and SHUT OFF

APU, LH ENG and RH ENG,
FIRE PUSH .... Confirm and select

Water (or terrain) .... Contact
with minimum forward speed, but not less than stick shaker speed and at minimum sink rate.

When the airplane has stopped:

APU BOTTLE switch........................ Select, to discharge

Both engine BOTTLE switches......... Select, to discharge

Doors and overwing exits .... Open

Passenger evacuation .... Initiate

See “Passenger Evacuation” on page 8- 348.

Planned Ditching

NOTE:
This procedure is intended for use where sufficient time is available.

Descent.. Plan

Crew and Flight Attendants ....Alert and brief

ATC / Transponder ....Notify / Set 7700

ELT switch...ON

NO PED / SEAT BLTS switches ....ON

Loose equipment.................................... Secure and stow

Flight compartment door ....Unlocked

GND PROX WARN CB (1B14) ....Open

AUDIO WARNING switches (All)............. Confirm and DISABLE

Survival equipment.. Check

Life vest, harness and belts........................On and tighten

NOTE:
Life vests should be donned but not inflated until outside of the airplane.

NOTE:
Life vest light plugs should be removed only if ditching at night.

Shoulder harness.................................. Tight and locked

LDG GEAR lever .... UP

Prior to reducing speed below 145 KIAS:

FLAPS ....Use 45 degrees

CONTINUED ON NEXT PAGE

NOTE:
If on ADG power:

- The slats/flaps will operate at half speed:
- A momentary loss of ADG power may occur:
  - At 140 KIAS and below, if the slats/flaps are operating At 108 KIAS and below, if pitch trim is used.

NOTE:
If possible, ditch in the vicinity of rescue vessels, near coastlines or islands.

At approximately 2,000 feet:

Sea conditions and wind direction .... Determine

Ditching heading .... Establish

- Windspeed less than 15 knots: contact parallel to swells.
- Windspeed 15 to 45 knots: compromise between wind and swell.
- Windspeed greater than 45 knots, land into the wind.

Descent rate/approach speed .... Establish

L and R PACK switches .... Select off

PRESS CONT switch .... MAN

MAN ALT switch .... UP

MAN RATE knob .... Maximum INCR

When the airplane is completely depressurized:

BLEED VALVES selector .... CLSD

MAN ALT switch .... DN

FUEL, L and R BOOST PUMPS .... Confirm and select off

APU (if not required) .... Shut down

CONTINUED ON NEXT PAGE

At approximately 500 feet:

Crew and Flight Attendants .... Alert that ditching is imminent

Radio .... Transmit final position

Brace for landing .... Order on PA

EMER LTS switch .... ON

LANDING LTS switches (all) .... ON

Just Before Water Contact:

Thrust levers .... Confirm and SHUTOFF

APU, LH ENG and RH ENG,
FIRE PUSH .... Confirm and select

Water .... Contact with minimum forward speed, but not less than stick shaker speed, and at minimum sink rate.

After water contact, when the airplane has stopped:

APU BOTTLE switch .... Select, to discharge

Both engine BOTTLEs .... Select, to discharge

Doors and Overwing exits .... Open

Passenger evacuation .... Accomplish

See “Passenger Evacuation” on page 8- 348.

Planned Forced Landing

NOTE:
This procedure is intended for use where sufficient time is available.

Descent .... Plan

Crew and Flight Attendants .... Alert and brief

ATC / Transponder .... Notify / Set 7700

ELT switch .... ON

NO PED / SEAT BLTS switches .... ON

Loose equipment .... Secure and stow

Flight compartment door .... Unlocked

GND PROX WARN cb (1B14) .... Open

AUDIO WARNING switches (all) .... Confirm and DISABLE

Survival equipment .... Check

Harness and belts .... On and tighten

Shoulder harness .... Tight and locked

LDG GEAR lever .... As required

Prior to reducing speed below 145 KIAS:

FLAPS .... Use 45 degrees

NOTE:
If on ADG power:

- The slats/flaps will operate at half speed.
- A momentary loss of ADG power may occur:
  - At 140 KIAS and below, if the slats/flaps are operating. At 108 KIAS and below, if pitch trim is used.

CONTINUED ON NEXT PAGE

At approximately 2,000 feet:

Landing area conditions and wind direction.... Determine

Descent rate/approach speed.............................Establish

L and R PACK switches ....OFF

PRESS CONTROL..MAN

MAN ALT switch... UP

MAN RATE knob................................. Maximum INCR

When the airplane is completely depressurized:

BLEED VALVES switch ....CLSD

FUEL, L and R BOOST PUMPS ....Confirm and select off

APU (if not required)....................................... Shut down

At approximately 500 feet:

Crew and Flight Attendants .... Alert that forced landing is imminent

Radio..Transmit final position

Brace for landing ....Order on PA

EMER LTS switch...ON

LANDING LTS switches (all)....................................ON

Just Prior to Contact:

Thrust levers .... Confirm and SHUT OFF

APU, LH ENG and RH ENG,
FIRE PUSH.......................................Confirm and select

Terrain...Contact with minimum forward speed, but not less than stick shaker speed, and at minimum sink rate.

CONTINUED ON NEXT PAGE

When the airplane has stopped:

APU BOTTLE switch........................ Select, to discharge

Both engine BOTTLE switches......... Select, to discharge

Doors and Overwing exits ....Open

Passenger evacuation .... Accomplish

See "Passenger Evacuation" on page 8- 348.

Passenger Evacuation

Captain:

PARKING BRAKE .... On

Evacuation .... Command to FO

GND LIFT DUMPING switch .... MAN DISARM

Thrust levers ....SHUT OFF

Evacuation .... Initiate using PA system

APU, LH and RH ENG,
FIRE PUSH switches...Select

BATTERY MASTER switch ....OFF

First Officer (on evacuation command):

ATC ....Notify
of emergency conditions and of intention to evacuate

NOTE:
If ditching, disregard EMER DEPRESS action.

EMER DEPRESS switch...ON

EMER LTS switch................ON with PA announcement

Both pilots:

Appropriate exits...Open

Passenger evacuation .... Assist and direct passengers
away from airplane

Airplane ....Abandon
by any suitable exit

Emergency Equipment Locations

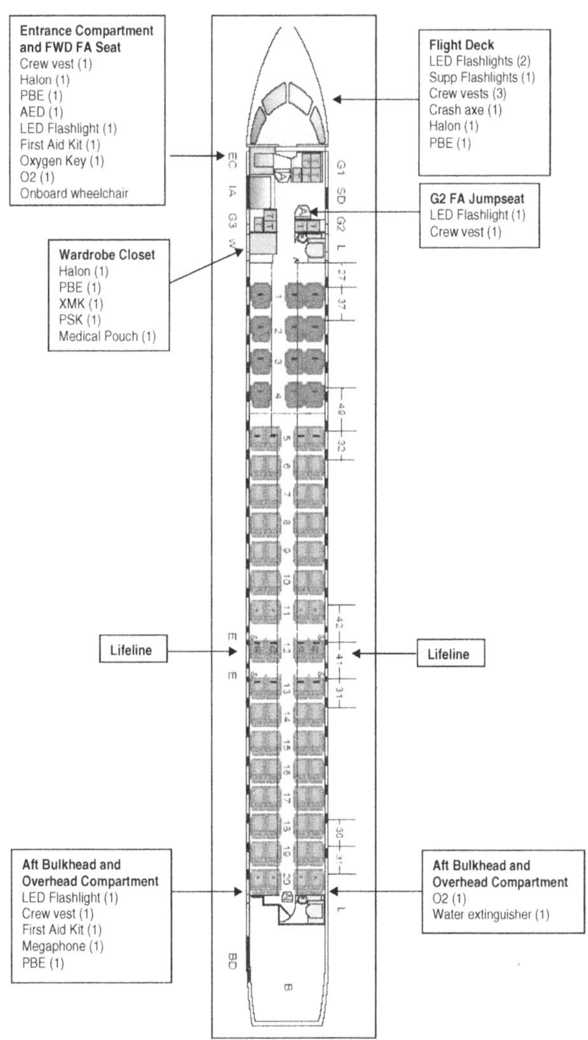

Index of References

Removed.

Intentionally Left Blank

Made in the USA
Monee, IL
26 August 2023